Powering India

Published by
ACADEMIC FOUNDATION
in association with

examine engage empower

Independent Power Producers
Association of India (IPPAI)

Powering India

A Decade of Policies and Regulation

Edited by

S.L. Rao

ACADEMIC FOUNDATION
NEW DELHI

First published in 2011
by

ACADEMIC FOUNDATION
4772-73 / 23 Bharat Ram Road, (23 Ansari Road),
Darya Ganj, New Delhi - 110 002 (India).
Phones : 23245001 / 02 / 03 / 04.
Fax : +91-11-23245005.
E-mail : books@academicfoundation.com
www. academicfoundation.com

in association with the
Independent Power Producers' Association of India (IPPAI)

Powering India: A Decade of Policies and Regulation
Edited by S.L. Rao

ISBN 13: 9788171889075
ISBN 10: 8171889077

Typeset by Italics India, New Delhi.
Printed and bound in India.

Contents

Section I

Overview

Section II

Nuclear

Section III

Renewable Energy

Section IV

Coal and Gas

Section V

Open Access Scenario

Section VI

Regulation: The Way Forward

About the Editor/ Contributors

Editor:

S.L. Rao has worked in multinational companies, was President, All India Management Association (1985); Executive Chairman, National Management Programme (1986-1988); Director-General, National Council of Applied Economic Research (1990-1996) and first Chairman of Central Electricity Regulatory Commission (1998-2001). He is presently a columnist in the *Telegraph* and *Deccan Herald*. He has 15 books to his name, on the economy, markets, energy, management, management education and governance, and innumerable articles. He is on many Boards of corporate, research, cultural and development organisations.

Contributors:

Bhanu Bhushan, Ex-Member, Central Electricity Regulatory Commission, played a pioneering role in establishing the Availability Based Tariff (ABT) regime, which has vastly improved the power system operations in India.

Gaurav Gandhi holds Bachelors in Mechanical Engineering from University of Pune, India and Masters in Renewable Energy Systems from University of Aachen, Germany. He is working with Suzlon Energy Ltd. in strategic business development (SBD) and is responsible for techno-commercial project validation, strategy and planning and policy and regulatory issues.

Balawant Joshi, Director, ABPS, is involved in Task XX (Branding of Energy Efficiency) and Task XXII (Energy Efficiency Portfolio Standards) of International Energy Agency Demand-Side Management (IEA-DSM) Programme.

Pradeep Kumar is a Chemical Engineering graduate and has Masters in Power Management. He is working with Suzlon Energy Ltd. and his areas of expertise are in policy and regulatory framework and structuring for commercialisation and market expansion of wind energy in India.

Anju Rani, advocate and legal consultant, specialises in advising interested parties on the infrastructure structure—public-private partnership (PPP) projects, concession agreements, and other corporate commercial transactions.

R.K. Sachdev is former Advisor (Coal) to Government of India, Ministry of Coal and President, Coal Preparation Society of India.

Krishna Sarma, Managing Partner, Corporate Law Group, counsels the State of Assam in the Supreme Court of India and is an Advocate-on-Record in India's Supreme Court.

Chintan Shah, Vice-President & Head, Strategic Business Development, Suzlon Energy Ltd. is an expert in policy and regulation that encompasses renewable energy projects, climate change and tariff structuring.

Ajay Shankar, Distinguished Fellow, The Energy & Resources Institute (TERI) was instrumental in initiating the dynamic Delhi-Mumbai Industrial Corridor Project for developing smart green cities.

Umesh Kumar Shukla possesses qualifications in finance, law, public policy and management and over 26 years of multi-disciplinary experience in different working environments. His research work has been published in India and abroad by IIM Bangalore, Emerald and Elsevier. He has presented many papers at national and international conferences organised by IIT-Kanpur, IIM-Bangalore and Syracuse University, USA.

Rupa Devi Singh, Managing Director & CEO, Power Exchange India Limited (PXIL), a pioneer of PXIL was instrumental in the setting up the first 'day-ahead' power exchange in the country.

Ashok Thampy is Associate Professor in Finance at the Indian Institute of Management, Bangalore (IIM-B). His areas of interest are corporate finance, financial risk management, banking, industrial economics and regulation.

Raghvendra 'Raghav' Upadhya is Senior Vice-President, Independent Power Producers Association of India (IPPAI). His areas of interest are energy policy, regulation and economics.

Foreword

India with a population of 1.1 billion people aspires to achieve double digit GDP growth to lift vast sections of its population out of poverty and also to create a modern, diversified industrial and services based economy. One of the most formidable challenges it faces is the lack of adequate infrastructure and the scarcity of funds to create new infrastructure.

China which had a similar amount of electricity generating capacity as India had 60 years ago, now has about 900,000 MW of power, compared to less than 190,000 MW in India. Most of this increase in China has come in the last 10 years, where the capacity addition was almost 100,000 MW per annum.

The challenges faced in the acquisition of land for setting up fresh power generating capacity, non-availability of legally enforceable fuel supply, and the uncertain and inconsistent policies and regulation particularly on allocation of coal blocks, and environmental clearances continue to deter investments in the power sector.

These have led to a rapidly growing deficit in India's ever-increasing demands for energy, especially that for electricity. Our losses on account of the power sector alone are expected to touch Rs 100,000 crore per annum.

While the government continues to withdraw from large parts of the infrastructure sector, making way for the private sector under various initiatives including public-private partnerships (PPPs), the policymaker and regulator remain unclear about the role of the private investor in the crucial, hitherto state-controlled sector.

The private sector looks to invest and make a reasonable profit. However, the flip-flop approach of state policymakers is ensuring that this does not happen. The investor feels that the policymakers have a very positive attitude at the outset but later on they let the project and policies drift, resulting in uncertainty that creates panic in the mind of investors who are saddled with capital-intensive, long-term, low-return, project investments which are illiquid and difficult to exit. This creates a ripple effect of uncertainty across the sectors.

The role of policymakers and independent regulators is primarily to create a level playing field between the various stakeholders which will attract private investments into the power sector, so that capacity can be built up and India can have the 'energy' it needs to feed its growing economy. However, this is yet to be achieved as the regulator has not got the full support of the policymakers who question the authority of the regulator in some cases, thereby defeating the very purpose for which the regulatory commissions were set up.

Clearly, to meet the growing aspirations of India's population in terms of employment and economic development, bottlenecks in the energy sector will have to be addressed.

As an independent body that provides a neutral platform for the discussion and examination of issues critical to the development of the private power sector in India, the Independent Power Producers Association of India (IPPAI) was set up in 1994, after the Government of India opened power generation to private players and invited developers to set up power projects in India.

Today, IPPAI provides a cerebral platform and an engaging interface between key stakeholders in the energy sector—consumers, power producers, regulators, policymakers, Discoms, as also thought leaders, economists, jurists, social reformers and the media to deliberate and possibly reconcile differing perspectives and worldviews on solutions to issues that inhibit the growth of the energy sector and the economy across the world.

Over the years, IPPAI has organised over 200 conferences, workshops and seminars in India and abroad where key policy, strategic, financial, legal, regulatory and technical issues in the power, oil & gas and allied sectors were examined.

A volume of relevant papers and articles are compiled for every conference of IPPAI and released at the event. The proceedings and outcome of IPPAI's conferences such as the Regulators & Policymakers

Retreat, are forwarded in the form of interactive DVDs and recommendations and to key decision-makers in the government, regulatory bodies and industry.

IPPAI takes pride in its initiatives that have contributed to achieving clarity in the way forward in helping reforms in the power sector, getting independent regulation to work more effectively, articulating the concerns of power producers and promoting competition and consumer choice.

Over the years, IPPAI has successfully highlighted issues that have hurt the power sector, such as high aggregate technical and commercial (AT&C) losses, denial of non-discriminatory open access and instances of the lack of sanctity of contract.

Our international initiatives include the first ever 'Regulatory Dialogue' between US and Indian regulators in Washington D.C. in 2010 under the FERC-CERC memorandum of understanding, a series of conferences in the US East and West Coasts on 'The Politics and Economics of Climate Change' and the 'Asia Energy Security Summit' in Colombo in 2011.

For the first time, IPPAI is bringing out a collection of selected papers presented at conferences held recently, to look in detail at the developments in the power sector in India in recent years. The papers in this book were selected and edited by Prof. S.L. Rao, the first chairman of the Central Electricity Regulatory Commission of India, an eminent economist and management expert, and a leading authority in the area of independent regulation and policy formulation in the energy sector.

I have no doubt that this book will prove relevant and useful to experts, professionals in the power sector, prospective investors, and to general readers interested in understanding the regulatory and policy environment that shapes the energy sector in India.

—Harry Dhaul,
Director-General,
IPPAI

Preface

Independent Power Producers Association of India (IPPAI) is 17 years old. It has over the years become an important forum for discussing and disseminating ideas relating to the power sector and its regulation. IPPAI over time has also broadened the scope of its concerns to place policy and regulation of power in the context of the larger social and economic concerns of India and the world. The highlight of IPPAI's annual events is the Regulators and Policymakers Retreat in July each year apart from many other seminars and conferences. These bring out the views of the important parties in the sector including government officials, regulators, power generation, transmission and distribution companies, trading companies, exchanges, non-governmental organisations working on the sector, journalists, scholars and others. Each of these events had a book of specially commissioned papers that are available to participants and others and a summary of findings is issued at the conclusion of each conference. They have had considerable influence on policy, regulatory practice and understanding of the issues in the power sector.

This book has many of the papers presented in the last two years at some of these conferences, brought up to date by the authors, and other papers that round off the subject. Considering the scarce literature on the infrastructure sectors in India despite their importance to the economy, and to the subject of independent regulation, we expect that this book will fill a very felt gap.

In the years since 1991, the Indian economy was opened to foreign investment, imports, private investments, less government restrictions, lower and simpler tax regimes in both direct and indirect taxation, and generally a more liberalised environment. However, the power sector did not receive a similar comprehensive set of reforms. They were piecemeal and are continuing. The power sector was opened to private investment in generation in 1991, transmission and distribution in 1998, and electricity regulatory commissions were centrally legislated in 1998. Work on a consolidated Act on electricity started soon thereafter and resulted in the Electricity Act, 2003.

The Orissa Electricity Regulatory Commission was created in 1994. After the Central legislation, the Central Electricity Regulatory Commission (CERC), state regulatory commissions for each state and union territory, and in the case of smaller geographies, joint commissions to cover more than one state, were created. Functions of the Central and state commissions are clearly defined. Since electricity is a concurrent subject under the Constitution, there is no reporting relationship of the state commissions with the Central commission. To overcome this and ease the operations of a national grid, a non-governmental Forum of Indian Regulators (FOIR) was created in which all regulators were members that would meet periodically to exchange ideas, determine directions, run training courses for regulators, and generally ensure a common approach to issues.

Almost invariably the commissions have had as chairman and members, retired officials from the IAS and sometimes from the other Central services, members of the Central Electricity Authority (CEA) of the Ministry of Power and state-owned electricity distribution enterprises, as well as generating companies. Under the High Courts and Supreme Court was formed a Central Appellate Tribunal for Electricity to hear appeals from the electricity commissions. These commissions are legislated to be independent, function transparently, in open hearings, consult all interested parties, and to give reasoned decisions. All documents available to them are usually accessible to anybody. They were expected to promote competition, encourage investment and safeguard the consumer interest.

The Electricity Act, 2003, introduced many new concepts that were not formally recognised earlier. It recognised the place of competition in electricity, the need to safeguard consumer interests and encourage investment in the sector. The new concepts included captive generation, electricity trading, electricity exchanges, open access to transmission and distribution wires, merchant power, tariff based bidding, and a hint of a Central role over state regulatory commissions by the creation of a Forum of Regulators of which the chairman would be the chairman of the CERC.

The Central government also developed policies that encouraged large plant capacities to be built by private as well as public enterprises. There were significant tax concessions given to mega and ultra mega power projects (UMPPs). With coal being a nationalised item, and denationalisation a political hot potato, government used the back door of coal linkages and captive coal mines as part of the tariff based competitive bidding process for such large capacities in the UMPPs. A thrust on renewables, particularly wind and solar energy, were supplemented by an innovative renewable energy certificate scheme.

After 1998 the Central government also tried to play a role in getting state governments to run viable state power systems, eliminate thefts, ensure metering of all customers, privatise distribution, and improve efficiencies at all levels in the state. An attempt was also made to bring about payments discipline by enabling securitisation of the payables of state electricity boards to Central government-owned undertakings.

The book serves three purposes: examine the functioning of independent regulation in all its aspects; study the role and feasibility of competition and enable an understanding of how the new concepts are being implemented and the ways in which they can be used.

The writers are all experts in their chosen areas. I am obliged to them for permitting to reproduce their papers. I am also grateful to Raghavendra 'Raghav' Upadhya for his unfailing good humour in helping me, Iqbal Kaur for her follow-up, Sanu Kapila of Academic Foundation (Publishers) for taking on this book at short notice, and to IPPAI that made the book possible.

— S.L. Rao,
Former Chairman,
CERC

Section I

Overview

1

The Role of Independent Regulation in India

S.L. Rao

Independent regulation is a form of governance that invites participation by all concerned, is transparent and responds to all arguments in giving its decisions. This is in sharp contrast to our bureaucratic and procedure bound governments in India. However, there are weaknesses in the way in which this new institution that has been introduced in some sectors is being implemented. I shall focus on the regulation of infrastructure and how it might develop by 2025. In this, I shall touch upon the problems, the context and the issues.

The Problems Today

Governments in India are tied down by complicated procedures and processes. They make decision-making slow and convoluted, and delivery of government services of all kinds inefficient, wasteful and corrupt. The lack of accountability among individual decision-makers prevents any change for the better except for the rare enlightened individual government officer. Development, especially in physical (and social) infrastructure is slow, running far behind needs, poor in execution and hesitant in allowing private participation. Dominance of government-owned and-run enterprises makes for secretive and collusive decision-making. Private entry is slow because of uncertainty about fairness of government decision-making procedures.

The main decisions relate to licensing, tariffs and quality. In physical infrastructure, investments are large and lumpy, with long gestation periods and limitations on what tariffs can be charged. Government departments are biased in favour of the public sector: for example there was a nine-year holdup after legislation was passed to permit it, of private entry into electricity transmission; protecting the monopoly of the state monopoly company; interference by state governments in tariff fixation; use of 'revenue share' as the decider of tariffs for private terminals in major ports, thus benefiting the port trusts but not making the tariffs user-friendly.

If infrastructure is to grow, attract investment, and meet needs, all concerned parties must be consulted. Decisions must be made speedily and reasons given for decisions, dealing also with any objections that might have been made in the process of consultation. Everything must be visible, including all documents considered for the decision. These considerations should rule all government decisions so that there is a sense of participation against the present sense of alienation of people from government. Government departments do not make decisions in a transparent manner. Opaqueness, lack of consultation and reasoning, has marked most government decisions on such matters in infrastructure.

The independent regulatory institutions that have been created have helped, particularly in telecom where private entry and technological changes have vastly expanded the market, lowered tariffs and improved quality. The Department of Telecommunications (DoT) was compelled by private investors and public opinion to distance itself and permit rational decision-making.

Electricity is the other major sector that has had independent regulation nationally and in almost all states since the late 1990s. While this has improved transparency and largely eliminated the bias of Central government in favour of central enterprises in tariff decisions, it has not enthused private entrants. Some instances: state regulators have delayed justifiable tariff increases, parked legitimate expenses as 'regulatory assets' to avoid tariff increases at the behest of state governments, held up electricity trading by placing a very low cap on trading margins, delayed open access on transmission lines by fixing very high surcharges.

While legislation and government policies have tried to ease private entry into electricity, implementation has been delayed or sabotaged by ideological commitment to public enterprises both by the bureaucracy and by regulators.

Among the reasons why independent regulation has not performed as expected in electricity sector, is the lack of conviction of politicians in power and senior bureaucrats in the new culture of openness, participation and reasoned decisions that independent regulation was expected to introduce. State governments have many times subverted regulatory decisions, tried to control regulatory decisions, while the Central government has condoned their public enterprises that do not implement the law or engage in time consuming litigation against regulatory orders that are then upheld by the highest courts. Further, those appointed to regulatory bodies have many times brought their baggage of experience in a 'command and control' economy.

Many other sectors regulated by government also face the problems of opacity, corruption, lack of monitoring and inspection, and poor service to the user. Some examples are professional and technical education, pharmaceuticals, hospitals and nursing homes, coal, oil and gas, roads, railways, urban infrastructure, airports and civil aviation.

Issues

In creating independent regulatory bodies, central and state governments appear to have acted for reasons other than a desire to improve speed and transparency in decision-making, better service to users and stimulating substantial private investment. The attraction was not the transparency, open consultation, speedy and reasoned decisions that independent regulation could bring about. It was the post-retirement senior appointments in such bodies that have been and will become available in such independent regulatory bodies, usually to retiring government servants. Almost every chairman and member of regulatory bodies is a retired government servant. It is not surprising that almost every Central government department, many state governments and even local bodies, talk of setting up independent regulatory commissions!

If independent regulators are properly selected, their decisions respected and implemented, and they function as they should, they can bring considerable improvement. They can correct the inadequate if not poor involvement and participation by consumers and other concerned interests. They can simplify and speed up decision-making, affected today by government procedures and processes that are complex and secretive. They could be immune to political and other influences that prevent faithful implementation of programmes. They will unlike governments, give reasoned arguments to show why one decision was made as against other options. The regulatory decisions could over time develop into precedents and case law.

One important bottleneck in Indian governance that prevents coordinated decision-making even by independent regulators as in electricity is that there is no holistic decision-making, only sectoral (for example, oil and gas prices are decided without reference to their effect on generation of power and production of fertilisers). This must change.

Infrastructure Today

Almost all infrastructures today are dominated by government ownership. This results in government enterprises acting in concert with government departments to protect their turfs. Yet, there is a growing presence of the private sector in telecom and in oil and gas, and to a much lesser extent in power. Minor ports have been set up by the private sector and run efficiently (without independent regulation). Major ports are inefficient except where private terminals have been allowed. The road sector and airports have introduced 'viability gap funding' and land development rights that have enabled the beginning of private sector participation. Water has seen no private entry except by water bottlers. Experience of independent regulation has been that private entities are easier to regulate and government-owned enterprises are much more difficult.

Civil aviation is in a mess, because of proliferation of airlines, poor airport management and rampant penetration pricing by far more airlines than available capacity for them at airports. Similarly road development is slow while traffic grows by leaps and bounds. Other sectors like education and health show little transparency, poor standards, low quality, and exploitation of parents and students. Pharmaceutical pricing is unsatisfactory to both producers and consumers, there is widespread production and distribution of fake drugs, quality standards are poor, clinical trials are many times 'fixed' and there is little protection to patients. All of them would benefit by better governance that truly independent regulation can bring.

There is little holistic decision-making, for example, with irrigation, drinking water and ground water usage, each of which is considered individually. This is so in other sectors as well.

In the case of energy, the use of commercial energy is still low and creates problems of pollution and disease. It is of importance to the poor and is practically unregulated and definitely not in the interests of the poor users. Many households are unconnected to electricity. Very many cannot afford market rates for power, kerosene and liquefied petroleum gas (LPG). Government supplies some users at low rates and sometimes free. Inefficiencies abound in production/generation, transportation/ transmission, distribution and supply. The state governments that are the principal owners of power generation and distribution are hard pressed to find resources for other governmental responsibilities since they are spending so much of state resources on supplying cheaper power. There is little attempt to look at the relationship of each energy resource to others and to users.

Today, the independent regulator exists in power, oil and gas and telecom. There is also one for tariffs in major ports but the agency can hardly be considered independent, being under the control of the Ministry. The power regulators have focussed chiefly on improving efficiencies, determining tariffs and improving quality. Independent regulation has not rationalised distortions between user groups in tariffs. There is modest improvement in transmission and distribution (T&D) losses and significant improvement in plant load factors in generation. Metering has expanded though it is still only for a portion of users. However, there is little change in using environmentally benign energy, estimating costs and usage by different consumer groups, targeting subsidies and keeping them in check, in moving towards cost of service for user charges, competition at least in serving bulk users, removing major transmission constraints, significantly reducing voltage fluctuations, improving safety, cutting technical losses in T&D, improving collections, imbibing professional management, commercial and entrepreneurial attitudes and mindsets.

The oil and gas regulatory board created in 2007 is a toothless animal, with no interaction with other energy sources, no tariff setting powers and is unconnected to the interests of the major users, namely power and fertilisers. It does not deal with problems caused by subsidies and cross-subsidies and the lack of adequate pipeline capacity. Coal remains a government monopoly and tariffs and quality pose major costs for users. The task in the coming years is to correct these conditions.

In these sectors the question of subsidising the poor and vulnerable groups has tended to be at the cost of the better-off customers. Subsidies are also reimbursed by governments, but what has been paid has been delayed, causing serious cash problems for the enterprises.

Environmental issues are impacted greatly by infrastructure sectors like power, oil and gas, coal and water. However there is no independent regulation of environmental concerns in these sectors, and the existing regulators have no mandate to do so. There is no coordination between different regulators and the environment approving agencies. Nor is civil aviation or roads or railways independently regulated for tariffs, operations and safety.

Accountability is an issue that needs to be addressed. There are instances of regulators in electricity, all of whose decisions have been overturned, and there is no one to pull them up. Others have been said to be too influenced by some of the enterprises they regulate. Some regulators have poor understanding of complex financial issues. Almost all issue orders but do not have regular monitoring to ensure implementation.

Infrastructure: A 2025 Scenario

1. By 2025 we could expect independent regulation to exist in many physical and social infrastructure sectors. Education and training in independent regulation would improve the quality of such regulators and make them really independent in their decision-making. We will see many institutions offering such training and education.
2. The changes in remuneration, with the Sixth Pay Commission's imminent report, could enable selection from a wider group than merely retiring government servants as is the case now. Tenures of regulators might also be made fixed.
3. All regulatory bodies must be subject to oversight by the immediate appellate body (tribunal of High Court) while the appellate tribunals

should like the high courts be subject to oversight by the Supreme Court or the National Judicial Commission (NJC) if it is created by then.

4. There must be legal compulsion on all regulatory bodies to sit in common benches issues of common interest and concern. At the same time the regulators must have much stronger penal powers to punish non-compliance.
5. Legislatures must oversee government ministries and demand that laws are complied with and implemented. At the same time, private sector presence in all sectors will improve the effectiveness of regulation. For example, in power we could expect much more private entry into generation and transmission and privatisation of distribution. Electricity sector in India might have around 40 per cent in the private sector. More competition in setting up new capacities, improved efficiencies and quality would result.
6. Oil and gas should see more complete regulatory oversight over tariffs, pipelines, quality and distribution and coordination with other agencies regulating power, transport and fertilisers.
7. There will be more of the road sector under private operation and also some portion of the railways, especially some passenger and freight services. Civil aviation will consist largely of private airlines and privately operated airports. Independent regulation of these sectors should improve their efficiency and service quality.
8. Introducing rules similar to those affecting corporate governance of listed companies should cover institutions that provide services in education and health. Poor delivery of education and health services by government at present, will improve as regulators independent regulators lay down rules, monitor and inspect for compliance. This is weak today with regulatory bodies like All India Council for Technical Education (AICTE), Medical Council of India (MCI), Institute of Chartered Accountants of India (ICAI), etc.
9. Greater coordination will be brought about between the Central Electricity Regulatory Commission (CERC) in relation to the State Electricity Regulatory Commissions (SERCs) and with other regulatory bodies. There will also be closer consultation between the

ERCs, other infrastructure regulators, and concerned governments. Ideally there will be provision for common benches to be formed when the decision of one body impacts heavily on another—for example, prices of coal, oil and gas and electricity tariffs.

10. Regulatory bodies in drug control, education, etc., apart from other areas in physical infrastructure, are poorly funded. Funding for all regulatory bodies must be outside government influence, though they must be audited and reported.
11. Selection and appointment of members of regulatory bodies should be done by an independent appointments commission, with representation from political parties, operating enterprises, users, etc.
12. Regulators will involve themselves in improving the management, information and commercial cultures in electricity and other infrastructure enterprises.
13. Regulators must promote trading, markets and competition where possible. In developing markets they must ensure that tariffs are at reasonable and affordable levels, monitor quality, and ensure that competition is the basis for awarding contracts.
14. There might be a South Asian Energy Grid for Power. This will mean that countries have to surrender some sovereignty to regulators since electricity in a grid follows the laws of physics, not of national boundaries.
15. Hydro and nuclear power will have increased as will private participation in them. Distributed power will be common, using gas in small package generators as well as biomass gasifiers and solar technologies serving closed or isolated communities both in urban and rural India. State-owned entities will function independently.

Conclusion

Independent regulation has the potential for transforming governance in India. Ideally, all government decisions should be made transparently, consultative and reasoned. There must be coordination between all aspects that affect a decision and government departments and independent regulators must be required to see that there is such coordination.

Implementation of regulatory orders must be with all the force of law behind it. Selections and appointments must be completely above board. Accountability of independent regulators must be to a judicial body. Legislatures must ensure that government implements the law and must watch regulatory decisions to see that they respect the legislative intent.

2 The Regulatory Framework

S.L. Rao

Do Suppliers Welcome Competition?

It is the constant objective of every supplier to hold his customers to using his product or service. He tries to ensure this in a variety of ways. He might make his supply appear to be particularly suited for a specialised purpose that other suppliers are not able to do. He might offer special prices all the time, or for a special occasion or period. He might use distributors to whom he offers incentives to bring in more customers and for them to consume more. He might try to achieve highly flexible responses to changing situations so that he can capture customers and keep them. He would try to build his offering into a brand so that customers have a special perception about his supply, differentiating it from that of others. This branding not only helps him to build customer loyalty—but might also enable him to charge a premium that customers pay willingly because of the brand image.

All these actions are attempting to procure a monopoly situation for the supplier in relation to his customers, so that they remain tied to him. The difference between competition and monopoly is that under competition this relationship is temporary. It is possible for another supplier at any time to tempt the customer through a perceived better offering. In a competitive situation every supplier is therefore striving to make his offer the most desirable so that he increases the number of customers and the extent of their consumption while attracting new customers. Suppliers welcome competition because it gives them an opportunity to break into a market and to increase their share in it. They try to create a monopoly situation after entering the market so that customers are tied to their supply. The preference of any supplier would be to tie his customers to his supply without any choice so that they have nowhere else to go. It is in the absence of such an environment that suppliers have to devise marketing and brand building strategies to attract and keep customers.

In the days of industrial and import licensing, the government estimated required supply capacities and licensed producers and other suppliers accordingly. The result was that in fact the customer was exploited. The customer got poor quality and service and paid high prices. With the end of licensing the situation has transformed itself. The customer has choice and suppliers are competing to attract them, with price and other incentives, quality of supply and service and new product features.

A truly competitive situation compels suppliers to place the consumer interest at the core of his business.

An example of a preference for a monopoly position, even by private participants, in infrastructure is that of the new international airport. The same issues have come up at Bangalore, Mumbai and Delhi. The threat of such a new airport makes the present airport operator during the period when the new airport/terminal is not ready, to greatly improve the services offered and the quality of the amenities. But the new entrants make it a pre-condition that they would get a monopoly on traffic at their new airport/terminal and if another new airport is to be built (as in Delhi or Mumbai), that they would have the right of first refusal. The new entrant does not want competition. He wants a monopoly, and perhaps understandably so, given the huge investments involved which he wants to safeguard and assure a good return to his shareholders. In the area of infrastructure it is not only the public enterprise that has over the years sought a monopoly supply situation. The private entrant has also sought the same.

Electricity Licensees

The Electricity (Supply) Act, 1948 gave a monopoly and monopsony position to the state-owned electricity boards. Within the state, they had the monopoly over electricity lines for transmission, distribution and supply except for the then existing licensees. The lines over which electricity moved were a monopoly of the state for intra-state transmission and distribution and of the Central government for inter-state transmission. They were regarded as 'natural monopolies'. No supplier could sell directly to any customer and was compelled to sell only to the State Electricity Board (SEB), except again in the case of the pre-existing distribution licensees. The Electricity Act, 2003 changes this situation. 'Captive' generators can now sell to their members thus circumventing the veto power of state government electricity undertakings (and also private ones) over third party sale of electricity. Parallel transmission and distribution lines can now be established. These lines need not any longer be 'natural monopolies' and any investor can set them up. If there is a natural monopoly over transmission lines there will be open access to them. The investor will invest if there is sufficient economic justification and he can be sure of a fair market. It appears that he can now have this assurance.

Bombay appears to be a market for electricity where monopoly and competition are combined without regulatory intervention thus far. There are four electricity distributors in Bombay, the state-owned electricity board, the local authority-owned Brihanmumbai Electricity Supply & Transport (BEST) and two private licensees, TATA and Bombay Suburban Electricity Supply (BSES). The state board is the default supplier to all of them and charges a hefty sum for that position. TATA is a 'sole' supplier to the other two and is also the owner of the transmission gateway to Bombay that is the only one available to them for any supplies from outside the state. At the same time customers are said to have been wooed away with special prices. This is possible when there are extra supplies in addition to that to be supplied under contract. The regulator has to decide whether such a monopoly/monopsony is permissible and that too with the new Electricity Act, 2003 opening up generation, transmission and distribution. Can any supplier in the meantime, so long as the old rules apply, poach on the customers of another with price offers that are not available to the distributor who also has to buy from him?

Clearly even the private licensee would like a monopoly or a competition when it suits him. The purpose of market regulation is to ensure that these individual preferences are set aside and common rules and principles are made to apply equally to all participants.

Privatisation of Distribution

Electricity distribution monopolies have eaten much of state government revenues and made them unable to perform their major tasks of governance. Commercially illiterate, unviable and technically inefficient SEBs ensured frequent power cuts and poor quality of supply. State ownership having proved incapable, privatisation (not competition) was seen as the answer. Orissa privatised distribution in 1997, Delhi in 2002 and Karnataka has been studying proposals for past several years. In Orissa, the sector remains heavily loss-making, tariffs have risen but not availability and quality. Delhi appears so far to be on track. In Karnataka only its secretive government officials, the World Bank and its consultants know the details of what it plans to do.

The Electricity Act, 2003 proposed fundamental changes to distribution through open access to all generators to the power lines that transmit and distribute electricity, freeing generation and ending state monopoly over distribution by allowing third-party sales. It has transferred to the Centre (the CERC and its chairman) many powers that were wielded by state governments. The Accelerated Power Development and Reform Programme (APDRP), securitisation of SEB debts to central public sector undertakings (CPSUs) and central policies to come on electricity, tariffs and fuels, will spur privatisation of distribution by state governments. The new Act will move high value and paying customers from SEBs to private suppliers. It has a five-year limit for eliminating cross-subsidies. It commits state governments to reimburse subsidies to SEBs. These will add to state government's burden unless they fully support new private distribution licensees.

The Electricity Act, 2003 has been in force for over eight years. Many of the expected changes have either not come about at all, or are happening with considerable dilatoriness. The state governments have not been in favour and have directed their regulatory commissions not to move on them. Thus cross-subsidies continue and have not been eliminated or even reduced; open access is beginning to be alowly introduced in some states; transmission and distribution losses remain high in most states; there has been no privatisation of distribution after Orissa and Delhi, both of which happened before the Act came into force. While the Montek Singh Ahluwalia Committee securitised over Rs 30,000 crore owing to central undertakings by state electricity enterprises, and payments to them are now more regular, the enterprises have continued to pile up losses because of uneconomic tariffs.

Orissa, an economically backward state with little agricultural electricity consumption, inefficient and ineffectively governed was ill-selected as the first to privatise. Private companies compounded it with poor 'due diligence'. Financial projections did not have agreed forward tariffs. T&D losses and thefts were guesstimates that proved wrong. The government pocketed gains on the sale of distribution companies. Huge accumulated liabilities were left with the single state-owned seller of electricity to the new private distribution companies, TRANSCO. The Orissa privatisation experience frightened off investors from further investment in state-owned distribution companies.

The Delhi privatisation modified the Orissa model but perpetuated the 'single buyer' model. But there was public and expert consultation that modified the proposals. Investor interest and requirements received much greater attention. Instead of guessing at T&D losses like Orissa, efficiency improvements were based on the gap between incoming quantity of electricity and revenues from sale. The state government laid down loss reduction targets and purchase bids had to be above them. The tariffs were set till 2006-07 and so that licensees would earn at least 16 per cent on paid-up capital and free reserves. The Delhi government would give Rs 2,600 crore as loan to TRANSCO that would supply electricity to distributors. This was to bridge the gap between TRANSCO's selling price and what distributors could pay after T&D losses (including thefts), and targeted collections. Until these leakages came under control on a loss reduction programme for each year for five years, the distributors would be compensated on an agreed schedule. There was therefore pressure on government also to help with law and order and other support.

The Delhi privatisation is believed to have saved the Delhi government over Rs 25,000 crore that they would otherwise have lost, on past trends. The T&D losses have halved. The supply has improved, also the demand has gone up further. The monopoly on procurement by the state government has been a disadvantage since the distribution enterprises would have procured electricity on long-term contracts. The regulator has made the enterprises unviable by not allowing legitimate expenses. New electronic meters, better maintenance and replacement of transformers and other equipment, use of police to catch and punish power thieves and the reduction in collusion in theft by employees, have all helped. However the Central government has not used the Delhi experience to push other states for privatisation.

Karnataka was close to privatising distribution. It has a lot more subsidised electricity supplies to agriculture than Delhi or Orissa. It also has a single buyer, a TRANSCO who will sell to distributors. But it did not propose to auction distribution companies on T&D loss reduction. It assured distributors a distribution margin to be made up by government. The investor was to be protected from tariff risks, collection risks and commercial loss risks in cases where the regulator or the administration is responsible. The investor takes the risks of collection in relation to metered consumers, reducing thefts below the levels established by experience, operations risks, risks in capital expenditure management and technical losses in distribution. The subsidy was for the financial deficit in the procuring and selling of electricity and to be made up by government. Each licensee would have a transition period after which he would pay all costs subject to specified timely actions by governments and the regulator. The State was to commit to doing many things to avoid further financial burdens. The Karnataka Electricity Regulatory Commission (KERC) would determine tariffs on a multi-year basis and was not to question the distribution companies on their numbers during a transition period that might vary from five to eight years. KERC criticised the scheme as based on poor data on baseline conditions, the possibility of complex litigation and for suspending the powers of the regulator. Ultimately the scheme was allowed to die. Since then the situation has deteriorated with mounting T&D losses, load shedding, large financial losses, compounded by free electricity to agricutlure, government pressures on the KERC to keep tariff increases to a minimum, mounting dues to state-owned generating companies and reduced operational efficiencies due to poor maintenance of plants for political reasons.

Investors do need assurances of government support to reduce thefts, install meters, collect payments for bills and improve efficiencies and guaranteed minimum returns. Low tariffs, subsidies and cross-subsidies have diluted investor interest in privatisation. The new Electricity Act must also be factored in.

However, the privatisation proposals so far have been much more focussed on reducing the future liabilities of state governments on account of losses in the SEBs than on improving the service to customers that competition and choice could bring about. The new Act enables this to begin.

Trading

Trading would enable supply and demand to be better matched. It requires a well-oiled enabling environment of regulatory institutions and practices and infrastructure. Adequate transmission, strong and objective load despatch centres, online accounting mechanisms, ability to correctly forecast supply and demand over short intervals, and many other conditions to be fulfilled.

The experiences in California with power trading have made some lessons clear for electricity trading. Electricity cannot be opened up in one go to match trading in financial and commodity markets with instruments like options, hedging, etc. Trading in electricity must always reflect the essential condition that base supplies must be kept going and tariffs must not be volatile. Hence the first condition of trading is that the 'asset light' philosophy developed by Enron cannot enter electricity trading.

Generation or financial assets must back all trading. These assets must account for a significant portion of the traded quantities. The trades must be genuine and not speculative in nature. Only excess supplies must be sold on short-term sales after meeting the contracted loads. The purely financial player must have control over generation capacities so that his trading follows similar rules.

What skills must a generator-cum-trader need if he is to make money? His skills would be in fuel purchases, generating efficiencies, accurate forecasting of supplies and demands, their geographic location and availability of transmission capacity when required.

After the Enron fiasco, the trend in the USA now is for financial institutions to support trading operations as counterpart to their existing financing operations. For example the Bank of America has entered into power trading of wholesale energy contracts on behalf of those in energy and utility sectors. It buys and sells financial contracts but avoids getting involved in the actual delivery of power. Even this is not desirable for us at this stage and we should restrict trading to those companies that have the support of generation capacities.

Telecom

The telecom sector has grown enormously after the advent of private entrants. At the outset government tried to create regulated monopolies or oligopolies and in the process tried to break telecom into tightly segregated sectors such as cellular, basic, long distance, international, etc. It took time to realise that for the customer, all of them were part of his overall need for communication. He would prefer to have them available to him seamlessly. The drivers of customer loyalty and differentiators between suppliers would be technology, price, quality and range of services, brand images created by marketing and advertising.

Independent Regulation of Electricity: Impact and Lessons

Background

The CERC and the SERCs were established with no local support groups lobbying for their creation. They did not arise out of academic research and discussion. Neither the electricity sector, largely Central or state government-owned with a few private licensees, nor consumer groups asked for the creation of independent regulatory commissions. Their creation was in response to the carrots from two external agencies, the World Bank and the Power Finance Corporation Ltd. (of India). The Central government was convinced that neither the Central nor state governments, nor indeed private investors in India would be able to raise the large sums required to improve quantity of quality electricity in the country. Thus there was no strong political or administrative commitment to their establishment except among some officials and ministers in the Central government. This absence of, or weak commitment to, the handing over of government powers to ERCs continues in all state governments and even perhaps at the Centre. The subsequent rocky relationships between many state governments and SERCs can be traced back to this imposition of the idea of ERCs from outside and not as a natural development arising out of local policy.

Policy and Regulation

Regulation is the inevitable component of any governmental and executive function. The legislative body enshrines the intentions and desires of the elected representatives through legislation. Government in addition also formulates meta-policies that are shaped by the ideologies and concerns of the parties in power. (For example the Common Minimum Action Programme of 1996 or the Accelerated Power Development and Reform Programme of 2000 were policy initiatives of the Central government that did not have to be part of any legislation and were not debated and passed by Parliament, but yet were important policy initiatives). These constitute public policy.

Legislation and meta-policies have to be translated into action. This requires the formulation of regulations that consist of rules and principles that have to be followed by all concerned. They are what give flesh to the bare bones of legislation and policy. They are essential for bringing the law to life as a series of actions.

In some countries the legislation is so drafted that there is less scope for interpretation by the executing officers. In India, as in the USA, legislation is so drafted that a fair amount of discretion exists with all executing officers to interpret the legislative intent especially in local circumstances. This is despite rules and regulations having to be framed for giving effect to the legislation. In some Latin American countries, the drafting is much more detailed thus limiting the scope for discretionary interpretation.

When the law transfers regulatory responsibility to an ERC it seems logical that, as it would with a government officer who formulates rules and regulations to implement the law, government must transfer the same powers to the ERC. These powers enable the government regulator or the ERC to interpret the law. To an extent all bureaucrats have such powers of discretionary interpretation of the law. These are also policymaking powers to the extent that they give scope for interpreting the legislation. Such powers must pass to the ERCs with the functions that are transferred to them from the executive.

Governments (especially in the States) having created ERCs now fear that ERC's might prevent reform in directions desired by investors, funding agencies and by governments. Some governments are advocating changing the law so that regulators can be bypassed. Overruling the ERCs will make them ineffective.

Legislation and broad public policy are the responsibility of governments and not of appointed authorities like the ERCs. There will be occasions when the public or national interest might require a direction that does not fit the logical reasoning of a quasi-judicial body like the ERC. The regulator despite hearing everyone may not appreciate the required balance between parties and might be even handed as he is expected to be between them, when a tilt in one direction is needed. Hence governments must retain the power to issue policy directives but must use the power with circumspection. The Electricity Act, 2003 expects the Central government to issue national policies that are binding on regulators, and Central and state governments have powers to issue policy directives to the ERCs at their desire on any issue. The Central government has never used this power to issue directives to the CERC. However state governments have done so, though many have done so verbally and not in writing.

Why Independent Regulation in Infrastructure

ERCs were expected to make private investors more comfortable about investing since they could expect that there would be no bias of pro-government attitudes towards the largely state-owned electricity sector. ERCs are supposed to take decisions more speedily than government has tended to do. ERCs are expected to bring about predictability. The law expects them to encourage investment in the sector while safeguarding the consumer interest. They were to promote competition, efficiency and economy in the sector. They are expected to function transparently and are quasi-judicial bodies. It was expected that they would improve the commercial viability of the sector.

These objectives have mostly not been achieved. For example the SERCs could not give certainty and predictability in tariffs because the

SERCs were able to decide tariffs a year at a time and not for three or more years forward. The reason was that the baseline information was faulty and inadequate, whether about assets, transmission and distribution losses, agricultural consumption, extent of metering or other matters. It was expected that at least the central sector would be better in this respect.

But even CERC could not get essential operating data from a Central government-owned company, National Thermal Power Corporation (NTPC), for reviewing operating norms set 10 years earlier and required to be reviewed after five years (though CERC made major changes despite this in the tariff structure in its jurisdiction). The company was shown to be making substantial additional profits as a result of these old norms on which they were getting tariffs. In the context of guaranteed returns and all costs as pass-throughs to be paid for by the customer, these extra profits should have gone to reduce tariffs to their customers, namely the SEBs.

The CERC and the SERCs were unable to improve on the information base. The lacuna affected the effectiveness of the SERCs much more than it did the CERC. Neither used their penal powers to demand the information. The SERCs did not commission their own surveys to establish baseline performance figures.

The CERC was able to function more speedily than the Central government used to do when it was in control of the regulation of CPSUs, etc. The SERCs were affected by their state governments who were at the 'coal-face' of electorates and vested interests and so were much more responsive to demands of SEB staff and to consumer protests. The pressure on the SERCs from many state governments (for example, Uttar Pradesh, Tamil Nadu, Andhra Pradesh) was quite considerable. The CERC had an easier job than the SERCs since it did not deal with retail tariffs in which politicians were more interested since they affected many voters.

In the 1980s and till 1998 the Central government played a key role in the determination of tariffs of electricity supplied to the SEBs from the central undertakings. This purchased power was on the rise, and became in due course the single fastest growing cost element in the SEB expenses. The central undertakings in generation and transmission were making growing profits while the profits of SEBs were declining and they were accumulating losses. The SEBs were unable to generate a positive cash flow also because of their populist pricing and inefficiencies, and also because they could not control power purchase costs (from Central government undertakings) and of other products and services also got from Central government-owned undertakings. Most of these decisions were taken with little consultation or concurrence from the customers, namely the SEBs. The main beneficiaries were the Central government-owned electricity undertakings and their owners—the Central government. The SEBs paid the costs of higher tariffs.

The CERC after public hearings reversed such unjustified impositions on the SEBs and corrected some of them to the disadvantage of the CPSUs. The Courts have so far upheld the CERC orders. The CERC was able to redress the balance between Centre and the states. It can do so long as it is independent, quasi-judicial and objective.

The states had milked the payers (industry) to benefit the non-payers (farmers and domestic consumers). SERCs were able, to an extent, to redress the balance between industrial consumers as against farmers and domestic consumers in the tariffs. Industry had been reeling from extortionate tariffs that boosted SEB revenues that became depressed by thefts and very low tariffs charged to farmers and domestic users.

The CERC and SERCs sought to redress these imbalances when they came out with their tariff philosophies and orders. The SERCs have also tried to be even handed. They have, along with other measures of governments been able to bring down the T&D losses and the financial losses in many SEBs, though they still remain at very high levels. However not all SERCs have been even-handed as was pointed out in a recent judgment of the Andhra Pradesh High Court that said 'the APERC appeared to be bending over backwards in favour of the AP state undertaking.'

Relevance of Independence

Some SERCs have (as pointed out in High Court judgments) favoured the interests of one party (usually the SEB) against another. Only a watchful public opinion and parties that have the courage to go in appeal can discipline the SERCs to be non-discriminatory.

Independence in action by the ERCs is expected to give confidence to investors that they will be treated with fairness unlike when the regulator was a government official and took decisions in private, without consultation and many times without giving reasons. But investors need much more than regulatory support. They need strong political commitment that will enable adequate tariffs, administrative support to curb theft and collusion, reimbursement of losses for an initial period when tariffs and thefts drive down revenues, support to deal with erring employees, etc. The ERC in the final analysis cannot be ahead of government on matters such as these. If government is unwilling to curb thefts, reimburse subsidy amounts to the distributing companies, or support remunerative tariffs, the ERC cannot be moving ahead of government.

Independence is not an end in itself but gives legitimacy to the regulator. Legitimacy requires capability on the part of the regulator to effectively discharge the functions allotted to him. This requires that regulators have adequate qualifications and experience, maturity, willingness to listen and adequate qualified staff that can study and advise on the issues. In the absence of this capability the legitimacy is in doubt and in that case independence is of no value.

The ERCs largely fail the test of legitimacy, as is evident, from a survey of regulatory commissions by Prayas, the TERI study for UNDP, analyses of Commission and Court rulings, and the relative lack of effectiveness of ERCs in achieving objectives. The new Electricity Act lays down the same procedures as its predecessor, and the search for members is to be conducted by eminent persons closely linked to government. The result of this procedure in the past has been that almost all chairmen and members of ERCs have been retired bureaucrats from the IAS who have spent their lifetimes in government service, along with other former government officials, others from the government electricity sector, and other government services like income tax, audit and such other services. Many of them are immensely capable and will not allow interference. Others bring the baggage of their past administrative or state electricity background, ideological orientation favouring state ownership, lack of commercial mindset and absence of professional management experience, which impede integrated approaches and fresh thinking.

Almost all staff members of ERCs are on deputation from other government departments to which they will revert after two or three years. Their professional skills are limited and not sharpened by widespread practice. For example their unfamiliarity with legal drafting may be a cause for many orders of some ERCs being inconsistent with each other and with those of other ERCs on the same issues. Nor does the appointment of expert consultants or advisors help to solve the problem. For consultants to be used effectively there has to be adequate capability within the ERCs to interact with and interpret the consultants.

There are questions therefore about the legitimacy of the ERCs that also reflect on their independence. These issues have never been addressed though the comments and criticism have been made since 1998.

But the principal reason for the relative ineffectiveness of ERCs would probably be the lack of strong political commitment, political will, and the absence of any strong conviction among Ministers and administrative officials about the need for independent ERCs and to help them be so.

Further, there is no compulsion on government and regulators to act strictly as required by the law. Thus government is to commence the search for a replacement of a member or chairman of an ERC six months before the vacancy arises and have him in position a month before the actual vacancy occurs. This requirement has rarely been followed as is clear from the Prayas survey. There is no consequence for not acting according to the law. Similarly there are ERCs that have not tabled annual reports, as required by the law, on the table of the concerned legislature. There must be some way in which the concerned government official and minister as well as the ERC are made aware that they must follow the law.

Though electricity is a concurrent subject in the Constitution, India has realised that optimum resource use requires a nationally integrated grid. Such a grid cannot be confined to geographical state boundaries. Even the USA, after the cascading blackouts over North America because of poor transmission facilities and limited national control over, the interconnections in the grid appears to have developed conviction among many that there must be a national grid. (However it has to face the stumbling block of powerful local private electricity interests in some states that push the state governments to oppose such a development on the ground that it is a violation of state rights).

A national grid, to function effectively, requires a number of enabling conditions. There must be effective regulation of the transmission system, independent load despatch centres, independent and non-profit regional transmission operators, real time information on all aspects of the power system, automatic under frequency relays to island load centres when necessary, automatic governor action to limit generation when necessary, good forecasting of loads and generation on a daily and short interval basis, strong discipline with severe penalties imposed by the regulator for violations, adequate, well-maintained and growing transmission capabilities, to name a few.

There is also need for a national Grid Code that is regularly reviewed and enforced. The CERC issued such a Code almost immediately after it was formed, and its regular review by the present Commission is to be commended. The effective functioning of a national Grid leads to growing central control over the system and the Electricity Act 2003 recognises the need for such Central control by the CERC.

In retrospect it appears as if the decision to make electricity into a concurrent subject under the Constitution did not recognise the immense benefits from an integrated national Grid (understandable because the technology for long distance transmission did not exist in 1951 when India became a Republic and adopted a Constitution). We now need to deliberate whether the states should continue to have the powers they still do on electricity or whether the Constitution should be amended to make it into a Central subject. Presumably the Central government realises that this will be almost impossible. Hence the new Act has used roundabout methods to overcome the dead weight of the SEBs; for example, in the very wide definition of 'captive generation'. A merchant generating plant can now be set up that will make third-party sales to its 'captive' customers. The end result is desirable but the subterfuge of making it 'captive' is necessitated by the Constitution.

With ERCs, Regional Electricity Boards (REBs), Regional Despatch Centers (RLDCs) and State Load Despatch Centers (SLDCs), real time information and accounting, there is a declining role for government departments in the Power system at the Centre and the states except in financially supporting vulnerable sections of the community. We need to find ways in which the need for separate ministries for power, coal, oil and gas, can be done away with and instead there are energy ministries that will enable better and coordinated development of national energy and fuel policies, energy security policies and coordinated tariff and investment policies.

Processes and Procedures for Transparency

Effective regulation of tariffs requires that the regulator is able to examine all aspects of costs. Today he cannot do so since fuels are under the control of other ministries and the regulator takes fuel prices as given. (It is another matter whether he actually needs to do so since he will be within his rights to demand that evidence that the fuel prices are competitive).

The investor requires the ability to satisfy his financiers about the viability of his power projects, and for this he needs to have certainity in electricity tariffs for a multi-year period. This has not been possible in the last few years, but the new Act requires SERCs to issue multi-year tariffs. The CERC has been given a role to play in enabling cohesiveness in approach by the SERCs in this connection. In this way, we will get a greater degree of coordination, consistency, cohesiveness and uniform speed in the introduction of new ideas and concepts in the electricity system. There is a lot to be said in the case of electricity for a strong centralised regulatory set up.

But is it also essential that a regulator should be able to scrutinise all the aspects that affect the subject of his regulation and not merely some aspect? When regulating electricity tariffs for example, in which fuels (coal, gas, naphtha, etc.,) account for the greatest part of the costs, is it reasonable to ask him to keep off scrutiny over whether their prices are justifiable because they come under other ministries, and there might be a separate regulator for them? This might protect ministerial turfs, but they prevent the regulator from being fully effective. Separate ministries have led to turf wars as Ministries and departments protect themselves from encroachment into their areas of responsibility. Each has its own vested interests, and would rather that each is dealt with on its own, separately. In that context would it be better to have narrow single sector regulators as for electricity today than for all related areas together as for energy? There is a strong case for rationalising our ministries and the related IRCs so that related areas can be considered together.

There is also a case for different regulatory bodies that impinge on one another to be directed to sit together, and formulate agreed principles so that such issues are taken into account. Thus, the Telecom Regulatory Authority of India (TRAI) and CERC (or concerned SERC in the case of a state) might sit together to consider what the telecom operator must pay to the transmission company for using its infrastructure for hanging his telephone wires. Or the electricity regulators might be directed to sit with the proposed coal, and the oil and gas regulators, to decide fuel prices. In the absence of regulators in such related areas, the ERC might sit with the concerned ministry or department.

The ERCs have followed a strictly judicial procedure. This needs modification when there are complex matters of principle and judgment involved that must be resolved in discussion, not through adversarial hearings. As provided in their Conduct of Business Rules (CBRs), ERCs should pursue a variety of methods in their consultation procedures, varying from the present solely public hearing process, as well as consultation papers to which written responses can be received and responded to, public consultations/conferences, and draft orders to which responses are received and studied before final orders are issued. Which method to use for which purpose will depend on the complexity of the issues to be decided, and the extent to which they fit into an adversarial model.

ERCs have achieved greater transparency and understanding of the weaknesses of the system. The lay public today better understands the significance of the T&D losses, and this is due to the public hearings before ERCs. The wide publicity given to the hearings at SERCs have brought about a better appreciation of the ills of the electricity system and the role of subsidies and cross-subsidies. Some SERCs have created mechanisms for hearing the lay public. The West Bengal High Court criticised this importance given to the consuming public but the Supreme Court has upheld it.

Different Models

Delhi and Uttaranchal state electricity Regulatory Commissions were one-member commissions. This had adverse effects on the quality of regulation since there was no tempering of one person's decisions by the views of others. However one of these commissions, after it became a three-member body, continued to announce very perverse decisions. After the retirement of the chairman, the members issued opinions quite contrary to what had been said by the chairman. These anomalies are a result of the poor political decision-making in having one-member commissions and in selecting inappropriate chairmen and members.

Of the commissions created in the last decade only the ERCs and Tariff Authority for Major Ports (TAMP) have tariff determining powers. The search, selection, appointment, termination process of chairmen and members is laid down for ERCs in the legislation, but not in the case of the others. Even for the ERCs there has been dilution in the new Act of the termination clause so that the termination of members is now by government instead of the head of the state. This process needs to be uniform for all Independent Regulatory Commission (IRCs) and must result in non-interference by the politicians, bureaucracy or the regulated entities. It has led in the case of the ERCs to selections having been almost entirely from among retiring bureaucrats and others from the electricity utilities. Staff has been mostly on deputation from other government departments to whom they will revert in due course. This does not give much scope for new thinking or a management, commercial and expert approach to matters.

There must be a wider population from which members and chairmen are selected. They must come with a variety of technical skills that have been honed in service outside government as well. For this, the remunerations paid cannot be confined to government limits for them, as well as for staff.

It has been said that even in the USA, such appointments are subject to nepotism and influence. For instance the chairman of the Federal Communications Commission (FCC) appointed by President George W. Bush is a young and relatively inexperienced man who is the son of the powerful Secretary of State in the US administration. But the US system which has only two political parties requires that such bodies have the majority, including the chairman, appointed by the ruling party, and the minority by the other party that is in opposition. Hence the appointments are entirely political, and the commission has to work in a manner that it moves forward despite the political nature of the appointments. It is not comparable to India where the commissions are expected to distance themselves from all such influences.

Except for the ERCs and TRAI, the other regulatory bodies are under the jurisdiction of the concerned ministry. We need to find an independent body to which they could be responsible in terms of meeting the reporting requirements laid down in the law.

Wherever the regulated entities have been largely state-owned, the task of the IRC has been more difficult. Wherever there has been a rapid change in ownership structure resulting in growing private entities, the task of the regulator has been easier. At least this is so in terms of compliance with orders. In the case of the ERCs the dominant government ownership makes one wonder whether the ERCs should, in the transition as the system gets more private operators, be more closely linked to government.

Commissions must all be multi-member and members must be full time, not part time or ex-officio (CERC has the chairman, Central Electricity Authority (CEA), as ex-officio member, who misses almost all hearings!). They must use a combination of processes and procedures to suit the problem that is being decided and not confine themselves to one standard procedure that is purely judicial.

But the secret of the effectiveness of ERCs seems to lie in the capability of regulators and their staff. This seems to depend on their proper selection (of regulators and their staffs). It also depends on the political commitment and administrative support that have been badly lacking at the state levels.

IRCs and Relationships

With Governments

The Central government has so far not issued any policy directives to the CERC. It has not asked the CERC, at any time, to make its orders go in a desired direction. It has behaved impeccably between 1998 and 2003 with the CERC. This is despite many ministers and different governments. The states have issued policy directives on tariffs and some other matters. Delhi government after careful deliberation issued a policy directive to the Delhi Electricity Regulatory Commission (DERC) to enable the distribution privatisation to proceed in the manner desired.

Governments might preferably not issue policy directives since they diminish the status of the ERC. Instead there must be a procedure for government to consult the commission, and get agreement with it on the policy than issue a unilateral directive. The alternative is to amend the legislation itself. This was done in the case of TRAI when in tariff rebalancing the authority raised local call tariffs *versus* those of long distance calls. Some people alleged that this would hurt the rural poor. TRAI was then reconstituted and the legislation amended so that it had no tariff setting powers and became largely a recommendatory body.

A major problem has been the inability of some state governments to meet their obligations to their regulated utilities in paying subsidies, providing law & order support, adequate staff to ERCs, etc. The SERCs have been unable in these cases to compel governments to do so. It is necessary for public opinion to take such instances to the higher courts for compelling the state governments to perform.

With Judiciary

The Supreme Court has upheld ERC's powers to consult widely, lay out principles and regulations as allowed under the law. The Andhra Pradesh High Court has criticised the Andhra Pradesh Electricity Regulatory Commission (APERC) for giving undue sympathy in its decisions to the state-owned entities. The Supreme Court has recognised that ERCs are created as expert bodies, and that therefore the courts should normally intervene with their orders only on matters of law, and not of fact or technical argument. Thus, the superior courts have been very accepting of the expertise of the ERCs and for such technical bodies to be created.

With PSUs

The ERCs at both Central and state levels have faced non-cooperation from SEBs and CPSUs in sharing correct data. In the states, the local utilities have frequently reverted with corrections to data already filed with the commissions. The implementation of orders of SERCs has been very tardy and poor. Even in the Centre there has been a deluge of review petitions and appeals that have held up implementation (of orders ultimately upheld) for years. Violation of SERC orders in the states has been quite common. The ERCs have been reluctant to use their penal powers and need to begin to selectively use them.

The Central government undertakings have a conflict with those of the state governments. At the Centre they initially thought that the CERC was only a government body and would therefore behave like the ministry had done in the past. It has taken many challenges in court for them to understand that there cannot be an adversarial relationship and that the decisions of the CERC as a quasi-judicial body could be changed only by the superior courts.

With Public

There are very few consumer groups in India with the desired degree of inter-disciplinary technical expertise in electricity, and this makes it difficult for them to appear, and present before ERCs. When they do, they tend to be more interested in large groups like residential and agricultural customers. They have had less focus on other issues than tariffs like electricity quality, merit order purchase and despatch, open access to enable better interchange of available power, trading and third-party sales. Now that the new Act makes these possible without the need for a hearing, the consumer groups might focus on such issues for quick actions. There is need to promote expertise amongst NGOs. Both ERCs and governments must provide funding to select NGOs so that expertise and capability can be developed among consumer groups.

Accessibility

There is a feeling among some top government officials that the SERCs could have shown a better balance between attending to the consumer interest, and encouraging the conditions for fresh investment into the sector. There has definitely been a tendency at least in some SERCs to favour the state-owned sector against private investors. This is reflected in the excessive charges imposed on the wheeling of captive power through the state-owned transmission system, or permissions tardily given or denied for setting up captive generation, low prices given to captive generated power sold to the SEBs, severe restrictions on third-party sales even when they had been earlier allowed by contract between the generator and the SEB, etc. Most of these measures, unfriendly to trading and private investment, will no longer be possible under the new Act. There will be a lot more consistency between the states.

But is it fair to blame the poor response by private investment on the ERCs alone and especially their 'consumer friendliness'? The ERCs are only one important measure, in a number, that are required to be taken by governments in order to attract private investment. There has been little demonstration of either political, or administrative commitment to support wholesale electricity reform in the states. The ERCs and their alleged imbalance against private investment and in favour of consumers cannot be the reason.

At the same time it is true that the actions of many SERCs do suggest a bias in favour of the state-owned distribution system. The past actions of the Central government were biased in favour of the central public sector (and the CERC redressed this). The reopening of signed power purchase agreements (PPAs) may be a stronger factor discouraging private investment since it raises doubts about whether signed contracts, on the basis of which financial closure was achieved, will be respected. We cannot take away the power of a government or a quasi-judicial body that approves contracts to reopen them if there were good reasons to suspect that they might have been entered through unfair means, or were exploitative. It is essential however that all ERCs lay down a common set of circumstances in which such reopening of PPAs might take place. As mentioned, the new Act can be expected to correct the bias in favour of the state-owned system.

Regulatory Law

It would have been correct to expect that after many years of functioning of the ERCs there was now a body of decisions of ERCs and the superior courts that amounted to a set of principles, and rules that all parties could refer to as part of the legal framework for electricity in addition to the legislation and announced government meta-policies. The judgments of the High Courts and the Supreme Court on appeals do build up to an evolving and fairly solid body of such regulatory law from the superior judiciary. This cannot be said of the orders of the SERCs on a number of important issues like captive generation, open access, third-party sale, etc., (Indeed the fact that these are now provided for in the new Act suggests that the SERCs were not getting it right.)

There are some inconsistencies and contradictions between SERCs and even between their own orders. They have made some difference in the determination of tariffs by redressing to some degree the imbalance between industry and others. They have brought awareness of efficiency issues and of theft of electricity. The broad direction is right in terms of effectiveness even if a consistent body of principles has not been evolved.

The courts have so far upheld CERC orders, but this is not the case with SERC orders. This may be because the CERC is dealing with broad macro issues on which considerable complex technical, engineering, economic, accounting and financial issues have to be considered together. The courts, by their own statements, prefer to leave such matters to the expertise of the special body created to consider such matters. In the case of the SERCs, the matters are much more contentious, and there are different ideological and other pressures that come into play. Different commissions have taken different positions on the same issues. It is not surprising that the courts have ruled against some of them. Yet, on some issues, there are visible directions to be seen from SERC orders as well. With the new Act many of these issues are now legislated for and there is no room for SERC interpretation.

Tariffs

Many SERCs have raised domestic and farm tariffs proportionately more than they have raised industrial tariffs. Considering that electricity tariffs to industry were so high that industries were finding them uneconomical and were making them uncompetitive, the direction of tariff decisions by SERCs has been good. It has been a good signal. If other enabling conditions that largely depend on government actions come into being this should encourage private investment in the sector. All SERCs have initiated steps in this direction.

Both CERC and SERCs have taken seriously the requirement to promote efficiency and have tried to implement performance based incentive regulation. However, they have not developed a strong base of information or a methodology (like the RPI-X by OFFER in the UK). There is no weightage to inflation, (though CERC did unscramble the wholesale price index and combine it with the consumer price index to develop a price index for inflation that was more suited to electricity). As a result, while targets have been set for T&D loss reduction for example by SERCs, they have never been met in any state. Many SERCs have nevertheless continued with setting targets on the same shifting sands of poor information. CERC and SERCs have not done anything to improve the baseline data, either by surveys initiated by them or by using their penal powers on the concerned utilities to provide the correct information.

No SERC has tackled the complex and politically sensitive issue of agricultural subsidies, for example, by capping them for each recipient, (in the absence of universal metering for example by capping the units for a sub-station) instead of giving farmers open-ended cheques to draw as much free or subsidised power as they wanted to.

I recall that Y.S. Rajasekhara Reddy won his first election to become Chief Minister of Andhra Pradesh, riding on the back of a promise to give free power to farmers. He kept his promise but this became a financial hemorrhage for the distribution enterprise in Andhra. I was one of those involved in a consultation at his invitation to suggest what could be done without breaking his promise to farmers. Suggestions included metering all farmers, ensuring their pump sets used electricity efficiently, regulating water use in each water basin, laying down crops in each basin so that high water intensity by drawing ground water did not reduce the water table, among others. Obviously none of these was acceptable politically and the state has continued to lose money in its electricity distribution enterprises.

While the commercial viability of some state distribution entities has improved, all of them still make substantial losses. SERCs by their actions alone can never turn them around. A great deal of government action is needed, from appointing managers, giving autonomy to management, laying down a policy on subsidies and then not interfering, reimbursing the entities for subsidies paid without delay, etc. These actions have not been taken. But in the case of tariffs the overall direction of many SERCs has been good, though not very successful in meeting efficiency objectives.

Captive Generation

Many states (Gujarat, Maharashtra and Rajasthan) have followed a restrictive policy by making it difficult and expensive to set up captive generating plants. In Andhra Pradesh, the APERC imposed high wheeling charges on captive power plants (CPPs) to make it economically unviable to set them up and operate them. They also tried to stop third-party sales allowed by earlier contracts. The High Court reversed the decisions of the commission. In Karnataka, both the government and the KERC allowed third-party sale, and made it easy to set up CPPs. In other instances, a seemingly liberal policy was obviated in the fine print. The new Act removes state discretion and SERC authority on these matters, and is far more liberal than anything done so far in any state.

Reopening Contracts

The Maharashtra Electricity Regulatory Commission (MERC), in effect, reopened the Dabhol-Enron PPA by revising demand forecasts downwards and asking the Maharashtra State Electricity Board (MSEB) to purchase power on merit order, that is, on the basis that the lowest cost power would be bought first. Since Dabhol was the most expensive because of extremely high fixed charges and high variables due to use of naphtha, MSEB had to stop buying from Dabhol. The original PPA had no indication that this could happen, and hence the action of the MERC could be said to be tantamount to reopening a signed PPA. On the other hand, the Godbole Committee that enquired into the PPA and related matters, established that the fixed charges were heavily padded, there was considerable opacity in purchase of naphtha and gas, and suspected other considerations having influenced the signing of the PPA than purely commercial ones. In such a situation would an ERC be justified in reopening signed contracts?

What is perhaps required is a clear statement of the exceptional circumstances in which contracts could be re-examined. This was done by the APERC who commenced by saying that they were not authorised to reopen signed contracts, unless there were renegotiation provisions, or the terms were onerous. The APERC then made a number of them infructuous. Other SERCs have not been so proactive.

Forum of Indian Regulators (FOIR) and CERC should study and develop a common set of principles that would be applied in considering whether a signed contract would be reopened by an ERC and make it known to all in advance.

Effectiveness

While some directions are right, the ERCs and especially SERCs have not been as effective as they might have. The ERCs were expected by the earlier Central Electricity Regulatory Commissions Act 1998 to promote competition, efficiency and economy in the sector, safeguard the consumer interest and encourage investment in the sector. None of these can be said to have happened. The ERCs of course did not have all the necessary powers for the purpose. The restrictions in the then prevailing legislation on third-party sales with veto powers with the SEBs, the lack of open access to the transmission system, the difficulty of entry and exit, restrictions on captive generation, are all factors that came in the way.

But some ERCs themselves encouraged many of these ills and even when they had powers, insisted on keeping the system closed to new entrants. While every SERC tried to bring about improvement in efficiency, all were rewarded with little success. They did not use the penal powers given to them to enforce better compliance with their orders. For fear of being accused of micro-management, they did little to set out detailed guidelines for implementation and milestones for monitoring. They did little to assure themselves whether their orders were capable of being implemented. Nor did they make any serious attempts to develop a better information base.

There has been little new private investment in the sector and little improvement in intra-state voltage, no improvement in commercial viability of utilities at state levels, or in the quality of supplies. Blackouts and brownouts are still common suggesting poor quality of power and poor maintenance of lines and equipment apart from poor availability.

But at the inter-state level, there is improvement in availability and significant improvement in frequency and voltage over the inter-state network over most of India. However, there has been little private investment in new privately financed inter-state generation capacity or in transmission.

SERCs have redressed the balance a little between tariffs to industry *versus* farmers and domestic consumers. But this has still a considerable way to go since cross-subsidies are quite substantial. The new Act requires cross-subsidies to be eliminated and it is expected that the Central government will soon either by amending the Act or through a policy statement, put time limits on achieving this and the rate of decline of cross-subsidies over the period.

The SERCs have not improved universal supply of electricity, primarily due to the poor financial state of the utilities, poor quality of lines and equipment, and lack of government support in raising tariffs and cutting losses.

There has been significant efficiency improvement at the Central level but not much in the states because of a lack of political and administrative support from governments and lack of significant authority with the ERCs.

Some SERCs have been consumer-friendly. KERC set up an Office for a Consumer Advocate. SERCs have encouraged people to appear before them. They have held hearings all over their states. But it is not clear what has been the achievement as a result.

Of course it has generated interest and more understanding about the problems of the sector and to that extent should over time help to carry people along on reforms in the sector. But it has not helped in voluntarily curbing thefts and collusion.

The CERC and the SERCs have done little to improve baseline information. The SERCs have not addressed the question of the neutrality of state load despatch centres that is necessary to implement the Availability Based Tariff at the state levels. It is now only in inter-state but has improved availability, quality and brought about trading. The SERCs also have to develop principles for the operation of true merit order despatch for which far more cost data is required than now available.

Trading and Markets

Ideally 'of course' the regulator should not be regulating tariffs but the markets that must be enabled to freely determine tariffs. The Electricity Act, 2003 takes us in this direction by creating the enabling environment for trading. However, a great part of electricity sales will inevitably be on long-term contracts since the base load has to be guaranteed with supplies. The Act is quite inadequate in going beyond direct trading by not providing for the developments of markets and an exchange.

Accountability

ERCs many times do not even follow the limited reporting to the legislature that is required of them. They favour some over others appearing before them. They cannot be disciplined.

Accountability is to the public opinion through the media. But a few other things are possible to make this accountability more precise and well defined.

An independent ratings process of the ERCs can be brought into being. As with ratings of other institutions (for example of business schools, companies, etc.), there needs to be more than one independent rating. It should not be by government.

Transition Directions

It will take at least a decade before the electricity sector in India has a large, and powerful private sector that can push for regulatory reform as the private sector is doing in the telecom sector. What should be the regulatory approach during the period of transition when the state-owned sector is dominant?

It seems apparent from the experience of the ERCs and particularly of SERCs that a loose federal approach will not work. It does not fit the requirements of electricity that follows the laws of physics and not of political boundaries. The power system requires up to the minute accounting, full metering, adequate transmission and distribution, good equipment maintenance, fail-safe systems to island load centres and cut load or generation as required by the system, strong grid discipline, close monitoring, efficient operations and highly professional management. It needs cohesiveness and consistency in policies. It needs strong centralised control and direction.

Conclusions

The infrastructure in India is largely state-owned, and certainly tightly controlled by governments. This makes the work of any IRC very difficult since the enterprises are close to the bureaucracy, and have their support for most of the challenge and defiance that they display against their orders. Since state ownership domination is unlikely to reduce in the near future we need to consider whether we can modify the system a little and retain the virtues of transparency and consultation while accepting that the regulators will be part of the government in a direct way.

It is apparent that we have too many state electricity regulatory commissions. They deliberate on the same issues and many times contradicting principles. Ideally we should have a central regulator with benches in each state. Alternatively, we should have regional regulators with the opportunity for each state to issue policy directives to them for their state on specified matters. It is possible that such changes to separate SERCs for each state might require changes in the Constitution of India. While that may be unlikely to take place there is the precedent of the ERC Act 1998 that was amended to permit joint commissions for small states. If that could be done perhaps it may not be so difficult to persuade states to accept the idea of regional commissions. It could save considerable expense. More importantly such bodies can build expertise from the limited people who are available.

A stronger role for the CERC in relation to the SERCs is highly desirable since the CERC has been better able to build and use expertise. The new Act gives the CERC a definite role in relation to SERCs, unlike the ERC Act, 1998 which gave it no role or relationship at all with SERCs.

We must also have a closer relationship between the ERCs and the concerned governments so that the ERCs march in step with the political and administrative commitment. This is also provided in the new Act though as pointed out earlier it goes against the principles of independence and transparency. Perhaps the minutes of the meetings of this new proposed forum could be publicised.

Another possibility is that we might consider the ERC as part of government but working in a transparent and consultative fashion with a fixed term for the regulators during which they cannot be moved. Their independence from government can also be written into the law so that increments, etc., are dealt with in a manner that there is no possibility that they could be used to make the regulators more amenable. Staff could be taken from outside government as well with less restrictive compensation policies than apply to normal government servants. The model of the chief election officers in the states has been described and discussed in the paper as a possible direction for development of this institution at least till such time that state ownership is dominant.

Competition in the infrastructure must be simulated since true competition is probably a long way away. This paper makes some suggestions in this respect.

Learning

What do these experiences of privatisation and competition teach us? The two are not the same. The customer and his satisfaction at a profit for the supplier must be at the core of all infrastructure service supplies.

There should be no monopolies and if they are required in a transition period, they must be so regulated that the customer interest is at the forefront in terms of availability, quality, access, affordability, etc. Any 'natural monopolies' must be made to ensure non-discriminatory access subject to publicly debated rules set out by a powerful regulator. Government must not impose monopoly/monopsony on any of the sectors even if it means that the public sector gets killed. The same applies to the private sector as in the example of the Bangalore airport. At the same time, private monopolies/monopsonies should also be eliminated. Here again a transition arrangement might be used before complete elimination. In order to achieve all this, a mechanism of independent regulation is essential that is speedy, objective, transparent and consultative.

3

Operational and Commercial Issues for New Power Plants

Bhanu Bhushan

The intent of this paper is to familiarise the reader with certain ground realities and special features of operation of electricity grids in India, as also with the tariff structure adopted for large thermal and hydro power plants. The discussion herein would be of interest to all those proposing to participate in generating capacity augmentation programme of India, as plant designers, equipment suppliers, plant owners/operators and investors. The issues covered include:

i) Overall power situation in India.
ii) Likely operating pattern of power plants.
iii) Must-run *versus* load-following for nuclear plants.
iv) Scheduling and despatch for nuclear plants.
v) Grid frequency.
vi) Regulation and frequency response.
vii) Grid disturbances, islanding and load rejection.
viii) Voltage fluctuations.
ix) Availability tariff.
x) Renewable energy.

Overall Power Situation in India

Reliable operation of the large interconnected grids of North America and Europe is founded on established practices of tight frequency control and all control areas sticking to their respective interchange schedules. The grid frequency normally remains within +/- 0.05 Hz of the rated frequency, and any excursion beyond that is considered alarming. Utilities' deviations from their schedules are minimal, and have to be made up in kind the next day. They are therefore not priced. Adequacy of generating capacity enables maintenance of requisite spinning and cold reserves at all times, for overcoming contingencies. In a regime with such discipline, all power plants must generate power according to the schedules decided by the concerned load despatch centres, and pit-head and nuclear power plants can steadily operate at a substantially constant MW as per their respective schedule.

The situation, on each of the above counts, has been very different in India. The peak-hour consumer demand far exceeds the available generating capacity. Capacity shortage is officially stated as around 15 per cent. Load-shedding is a daily routine except in metropolitan cities and state capitals. Rural supplies are regularly rostered commonly and restricted to 8-12 hours a day in most states. State utilities, in their anxiety or compulsion to minimise load-shedding in their area, tend to overdraw power from the larger grid. Interchange schedules go for a toss, and frequency often plunges below the stipulated lower limit. As per recent reports, the frequency is often below 49.5 Hz for over 10 per cent of the time in a day. On the other hand, industries and commercial establishments need back-up diesel generators for continued operation when power supply from the grid is cut-off or is curtailed (for a few hours every day), and domestic consumers have to bank on their own battery-backed 'inverters' to get the basic amenities of 'light and fan' round the clock.

Since deviation from drawal schedules of state utilities are inevitable and substantial, and cannot be returned in kind, they are priced. Utilities pay for over-drawal, and get paid for under-drawal at a frequency-linked rate, which goes up as frequency declines and goes down as frequency rises. This mechanism, known as Unscheduled Interchange or UI, was formally introduced at the inter-state level in India in 2002, along with the implementation of Availability Based Tariff (ABT) for Central government-owned thermal and hydro generating stations. It immediately succeeded in narrowing the frequency excursions from the earlier 48-51 Hz range to a more tolerable 49-50.5 Hz range. For the last two years, the Central Electricity Regulatory Commission (CERC) has been trying vigorously to 'tighten' the grid frequency further. Even so, the frequency remains within the prescribed 49.5-50.2 Hz range for only around 90 per cent of the time, still.

With continuing growth of population, rise of general prosperity, and Government's commitment to supply power to all, the electricity demand growth would continue to outpace generating capacity augmentation, and grid frequency regime may not improve much further for many years to come. The power plants to be set up in India shall therefore have to be engineered to operate in such frequency regime without too adverse an impact on their life and efficiency. Also, there would be a perpetual pressure on the plant operators to minimise the downtime, and maximise the power delivery at least during the peak load hours.

Likely Operating Pattern of Power Plants

The total consumer load in any system varies considerably over the 24-hour period. Even in industrialised countries, the night-time load is only about 60 per cent of the peak load in evening hours. Consumer load variation in a developing (less-industrialised) country like India, if it were not constrained by shortage of generating capacity, would be even higher. However, in our case, there is a considerable load shedding during peak-load hours and the peaks of daily load curves are severely chopped. On the other hand, agricultural loads are commonly rostered for enforced night-time running of irrigation pumps, which fills up the valley in daily load curves. As a result of the above, total system load varies over the 24-hour period to a much lesser extent, compared to developed countries. So much so, that most thermal and nuclear power plant operators in India take if for granted that their plants should be operating as base-load stations, even in case of load-centre stations.

The above position would, hopefully, change in the coming years as new generating capacity gets commissioned and the capacity shortage comes down, at least in percentage, if not in MW. The total system load should then see a larger variation from peak to off-peak hours, much larger than what can be accommodated by the hydro stations alone. While the large pit-head and nuclear plant could still be operating as base-load stations, all other plants (hydro, load-centre thermal and gas/liquid based) shall be required to back down substantially during off-peak hours. Many of the plants would even have to resort to two-shift operation, a mode which is presently put aside by the plant operators as unthinkable.

The deciding criterion as to which power plants would be on base-load, load-following or two-shift mode would be their variable costs, as reflected in their respective energy charge rates. In other words, all plants with a variable cost higher than that of the pit-head stations shall have to back down during off-peak hours. They would not only need to be engineered for such operation but their fuel supply system and contract shall have to be tailored for it. The latter would be particularly crucial for gas based generation and would require adequate 'line-pack' to be provided in the gas pipelines. Proposed increase in proportion of renewal energy would only aggravate the situation. In due course, thermal plants burning imported coal too may be required to back down during off-peak hours daily. It would be only expedient to have these plants engineered (in the beginning itself) to provide such backing down capability.

Must-Run *versus* Load-Following for Nuclear Plants

As of now, the aggregate generating capacity of nuclear power plants in India is only about 3 per cent of the total installed capacity. Being so small a portion of the total generation, the nuclear plants have been allowed to generate their entire available capability, and no backing down during off-peak hours has been insisted upon by their customers, more so in view of the prevailing situation of overall generation deficit and because of the nuclear stations claiming a special status.

As the percentage of nuclear plant capacity in the total installed capacity increases in future (as is being aimed at), the above aspect would need to be reviewed. It may then not be put under the carpet simply by raising the issues of operational problems, reactor safety and fuel rods' life. A much more pragmatic approach would be to bank on the low variable cost of the nuclear power generation, to get these commercial ventures to operate as base-load stations. In turn, this would require implementation of two-part tariff for nuclear plants, with clearly separated energy charge (to cover fuel and other variable costs) and capacity charge (to cover the fixed costs). As long as the energy charge rates of the nuclear plants are less than or of the same order as those of the coal-fired generation at pit-heads, the nuclear plants shall not be required to back down at any time. They would thus be running as must-run, without being categorised in that manner, and thus avoid the technical problems associated with load-following mode of operation, that too without any heart-burning.

The regional and state load despatch centers (RLDCs and SLDCs) already have a mandate for optimum scheduling and despatch, and have to ensure economic operation of the regional and state grids respectively. (Please see sections 28(3) and 32(2) of the Electricity Act, 2003.) They would automatically schedule the nuclear plants for base-load operation if their energy charge rates are lower than or of the same order as those of pit-head stations. Further, as long as the above criterion is satisfied, it is not necessary for energy charge rate to reflect the true variable cost of a nuclear plant, which need not even be disclosed except in case of regulator-specified cost-plus tariff.

The fixed cost recovery shall have to be effected through capacity charge, payment of which may be proportional or directly linked to the day-by-day MW capability of the plants. The capacity charge rate could even be specified in Rupees per MW per day, or some similar formulation. Such a tariff mechanism is already an established practice for thermal (fossil fuel-fired) generation in India and the same could easily be replicated for nuclear plants. It would equitably take care of the legitimate interests of the plant owners as well as their customers.

Much debate is taking place in other countries about load-following capability of nuclear plants, particularly where they constitute a large percentage of the total generating capacity. This would not be an issue in India, since nuclear capacity would remain below 10-15 per cent of the total installed capacity for many decades to come. As long as their variable costs remain well below those of load-centre coal and other fossil fuel-fired stations, nuclear plants would not have to do any load-following, i.e., they would not be required to back down during off-peak hours.

Let it be clarified here that by load-following discussed above, what is meant is the variation of generation over the day (from peak-load hours to off-peak) to match the variation in total system load hour-by-hour. Such duty is to be performed by thermal stations having higher fuel costs and by hydro-electric stations. With the foreseen generation mix, it is expected that nuclear and pit-head generating stations would always be scheduled to operate as base-load stations. However, they would still be required to provide primary frequency response and regulation (to be discussed later) to collectively shoulder the minute-to-minute fluctuations of system load and load-generation mismatch.

Scheduling and Despatch for Nuclear Plants

The nuclear power plants would have large capacities and generally be contracted to sell power to more than one state. They would therefore, fall within the jurisdiction of CERC, and would have to comply with the Indian Electricity Grid Code (IEGC), which can be accessed under 'Regulations' on the CERC website *(www.cercind.gov.in)*. Such plants shall be scheduled by the RLDCs according to Part-6 of the IEGC, directly relevant parts of which have been annexed (Annexure A-3.1) to this paper. A perusal of Sections 6.4 and 6.5 is recommended. It would be observed that India has opted for a scheme of decentralised control, in which generating stations are scheduled on day-ahead basis by the load despatch centres, but are permitted self-despatch depending on plant and system conditions. An important implication of this is that the plant operators need not be very precise or fussy about declaration of plant availability, and about generating power strictly as per schedule. Also, since deviations from schedule are permitted (though no longer encouraged in the new CERC approach), it is not necessary/mandatory to have the schedules revised every time the generation level changes or has to be changed, either an account of plant availability variation or due to system load-generation balance changes. It should however be noted that in the new CERC dispensation, over-generation and under-generation are not treated evenly. UI rate for the former is lower than the UI rate for the latter under certain conditions. If the generator is not alert, he can suffer financially, or be penalised for 'gaming'.

Grid Frequency

Frequency is the most crucial parameter in the operation of an AC system. The rated frequency in India is 50.0 Hz. While the frequency should ideally be close to the rated frequency all the time, it has been a serious problem in India. There was a time it varied from below 48.0 Hz to above 52.0 Hz, even beyond its legally permissible limit of +/- 3 per cent, i.e., from 48.5 Hz to 51.5 Hz as per Indian Electricity Rules, 1956. After introduction of Availability Based Tariff in 2002-03, the frequency has remained in 49.0-50.5 Hz band for most of the time. Attempts have been made since April 2009 to narrow down the band to 49.5-50.2 Hz, but as mentioned earlier, the frequency remained below 49.5 Hz for about 10 per cent of the time even in recent months.

Not being satisfied with the present frequency regime, CERC proposes to tighten it further, mainly by jacking up the penalties for deviations. Over-drawing utilities shall have to pay at even higher UI rates and under-drawing utilities will get compensated at lower UI rates. In effect, all deviations from schedules, whether good or bad for the larger system, are being discouraged. The long-term impact of these measures are yet to be seen, but their short-term impacts have been: (i) an inducement to purchase power (rather than overdraw as UI) through bilateral contracts or through power exchanges, (ii) consequent increase in price for traded power, (iii) increased load-shedding by utilities which are not able to purchase power, either due to short-supply or due to high cost and (iv) some improvement in frequency. It is the author's assessment that some of these developments are not healthy/positive, and the so-perceived improvements may not last too long, since the measures do not address the basic issue of generation inadequacy. Two typical frequency plots of these days are enclosed.

There are many reasons for such frequency regime, which would be considered unacceptable elsewhere. Suffice it to say here that India is unlikely to attain a reasonable frequency stability (of the order of +/- 0.1 Hz) even in the next decade, and grid frequency would continue to vary and fluctuate. The nuclear and other power plants set up in India would have to withstand such frequency variations and fluctuations, and therefore need to be engineered for trouble-free operation in such frequency regime. Specifically, the steam turbines must be capable of sustained operation in 49.0-50.5 Hz range, as is stipulated in IEC:45.

It has been mentioned earlier that the nuclear and pit-head stations would be scheduled as base-load stations and would be exempted from load-following duty, but would have to provide primary frequency response and regulation. What is basically meant by this is that if a generating unit is set to generate x1 MW, and frequency is constant at f1 Hz (f1 need not be equal to 50.0), it would be generating x1 MW, irrespective of whether f1 is higher than 50.0 or lower. When frequency changes from f1 to f2 Hz, and f2 is higher (lower) than f1, the generation would decrease (increase) from x1 to x2 MW without any intentional time delay, through free governor mode of operation (FGMO). However, the generating unit would not be required to continue generating x2 MW, but may return to x1 MW over the next few minutes, ramping up (down) at 1 per cent per minute or slower, even if the frequency stays at f2 Hz.

To explain further, the objective of regulation and frequency response in India would be only to minimise the frequency fluctuations through FGMO, but not to bring the frequency necessarily back to 50.0 Hz. Operation in the above fashion may require some reconfiguration of turbine speed-load control logic and a thermal buffer in the steam cycle to accommodate mismatch between reactor power/fuel injection and generator MW output of up to 5 per cent for a maximum duration of 5 minutes.

Why is Frequency Response and Regulation Necessary?

Frequency fluctuations are caused by load-generation imbalances in the system, and keep happening because consumer load keeps changing. Frequency fluctuations can be minimised if generation is changed in the appropriate direction without any time delay, which is possible only through FGMO. The present frequency regime in India is indicative of the generating units not providing this primary regulation and absence of frequency response through FGMO, due to which frequency fluctuates minute-to-minute much more than what it would if the generating units actually operated on free governor mode. For example, a 1 per cent change in system load (500 MW in a 50,000 MW system) would cause a frequency change of 0.3-0.4 Hz if no generating unit was on FGMO, and of only 0.02 Hz if all generating units operated on free governor mode.

Nuclear power plants are particularly susceptible to frequency fluctuations. As frequency changes, the speed of the coolant pumps changes proportionately, and the coolant flow and consequently the temperature differential across the reactor also vary. The above temperature differential is a primary signal for reactor power control, and its variation gives a command for change of reactor power even when the reactor has been steadily operating at the optimum level. This in turn causes unnecessary fluctuations of reactor power and undesirable wear of fuel rods, etc. Nuclear power plants in India have been affected over the years by the grid frequency fluctuations, but have still refused to consider operation on free governor mode, the only measure which can minimise frequency fluctuations. They have been able to get away from participation in this collective responsibility on account of: (i) nuclear capacity being a very small percentage of the total generating capacity, (ii) other generating stations also refusing to participate in frequency control and (iii) the nuclear stations claiming a 'holy cow' status. It is foreseen that all this would change in the coming years, and nuclear stations would also be required to provide primary frequency response and regulation as was outlined earlier, notwithstanding the fact that they have presently been exempted from this duty in the current grid code.

The conventional thermal generating units do not have a problem of the type described above, and can be set to deliver a constant MW. However if the grid frequency is to be stabilised, all conventional thermal units (which constitute the bulk of the generating capacity in India) must be on FGMO mode. The new super-critical units (with once-through boilers) would, however, have a problem similar to that of nuclear units. Frequency stability (free of fast fluctuations) would be equally critical for their well-being and they too must be on FGMO mode.

Grid Disturbances, Islanding and Load Rejection

There was a time when regional grids in India had very frequent grid disturbances, both minor and major. The nuclear and other plants then in operation had to bear the brunt of islanding and load rejection. Even tripping of generating units and poisoning out were not uncommon. The situation has considerably improved since implementation of ABT in 2002-2003. India has not suffered a grid collapse or a major grid disturbance in the last 8-9 years, though the risk of these still persists. It can now be said that the normal safeguards and provisions to withstand islanding and load rejection adopted for nuclear power plants in other countries should suffice in India as well. HP–LP bypass to dump the steam to condenser would generally be necessary for this purpose, along with capability to go on house-load and safe shut down in the event of a total disconnection from the grid or a complete grid collapse.

As for power plants of other types, they should be engineered for early restart once the station auxiliary supply has been restored from the grid. Thermal and renewable power plants generally need not have a self-restart or black-start capability, provided at least some strategically located hydro stations have this capability, and have the ability to supply auxiliary power to the thermal stations.

Voltage Fluctuations

While frequency is a system-wide parameter, and has one value throughout the grid on minute-to-minute basis, voltage differs from location to location. It is a local parameter depending on system and load configuration, and is required to be tackled locally through reactive power control. The grid voltage profile was a serious problem in India till 2002, but has since been effectively controlled through:

(i) Timely augmentation of extra-high valtage (EHV) transmission system, i.e., laying of additional transmission lines,

(ii) Installation of shunt reactors and series capacitors and

(iii) Installation of shunt capacitors in sub-transmission and distribution system.

The last item has largely been prompted by a scheme of pricing of reactive energy drawl by state utilities from the regional grids, details of which can be seen under Section 6.6. of the Indian Electricity Grid Code (IEGC), relevant extracts of which are enclosed in Annexure A-3.1.

The voltage profile on EHV grid has now improved to the extent that the large generating stations, which are generally away from load centres, are not required to supply reactive power for meeting the reactive component of consumer load. They have to generate or absorb reactive power mainly to maintain the voltage at the local EHV bus at the required levels, as per instructions of the regional load despatch centres. (Para 6 of Section 6.6 of IEGC may be seen in this connection).

It can reasonably be expected that, provided the necessary coordination with Central Electricity Authority (CEA) and Power Grid Corporation of India Ltd. (PGCIL) takes place sufficiently in advance, the requisite transmission system for the future nuclear and other power plants would be getting established in a planned and timely manner, and the plants would not face any special voltage profile problem. The standard and proven generator designs and excitation systems should therefore suffice. However, it may be desirable/necessary to provide on-load tap changers (OLTC) on generator transformers in order to maintain the generator terminal voltage at the optimum level under different operating conditions and requirements of reactive power generation/absorption.

Reactive power injection or absorption by generating stations under long-term capacity contracts and covered by ABT is not separately paid for as per the present philosophy in India. It is expected that the same approach would be adopted for the nuclear and other power plants of the future when contracted long-term for their capacity. However, for power plants not covered by long-term capacity contracts (e.g., renewable and merchant plants), reactive power injection/absorption may be priced according to the scheme described in Section 6.6 of the IEGC.

Availability Based Tariff

The Availability Based Tariff (ABT) successfully adopted in India since 2002-03 for the fossil fuel-fired central generating stations is described in Annexure A-3.2, a perusal of which is recommended. The nuclear power plants presently operating in India and owned by Central government have still not opted for its adoption due to certain apprehensions, which really are not well-founded. There is no valid reason why ABT concept should not be extended to the nuclear plants, to the benefit of all parties. Introduction of larger nuclear plants with single-part tariff (as presently practiced) is liable to lead to operational and commercial problems in grid operations, as was perpetually the case with Central thermal stations up to 2002. It would be advisable, particularly for private investors, to insist on adoption of ABT for dispute-free scheduling and operation of their generating stations, whether their tariff is determined on cost-plus basis or through competitive bidding.

For the guidance of those concerned/interested, relevant extracts of the CERC notification dated 19 January 2009 on terms and conditions of tariff for the April 2009–March 2014 period are attached as Annexure A-3.3. Please note that while ABT itself has only two parts, i.e., capacity charge and energy charge, its Indian version has a very important attachment, termed Unscheduled Interchange (UI) charge. This last component is for pricing of all deviations from schedules, both inadvertent and deliberate. As explained in Annexure A-3.2, the UI charge is at an energy rate linked to grid frequency. The rate goes up as frequency falls below 50.0 Hz (in a deficit situation), and goes down as frequency goes above 50.0 Hz (in a surplus situation). The UI rate as originally conceived and applied, is many things rolled together: pool price, incremental cost (system lambda), marginal price, balancing market price. UI is a multi-purpose tool which enables many things in India, a discussion on which would require a separate session. It is rather unfortunate that the present CERC dispensation, for details of which one should see the CERC regulation on UI issued on 30 March 2009 and amended on 28 April 2010, has moved away from the original simplicity of the scheme. One has now to be very cautious while trying to benefit from the mechanism.

Renewable Energy

Power plants generating energy from renewable sources (mini-hydro, wind and solar) have very different characteristics, and require a totally different treatment. Their special features, relevant in the present context, are as follows:

(i) Their level of generation is decided by nature, and cannot be controlled or regulated as per consumer load pattern.

(ii) While their generation may have a daily and seasonal pattern, any prediction of their hour-to-hour generation for the next day always has a degree of uncertainty. Therefore in case a renewable energy plant is scheduled day-ahead (for the next day) based on statistical generation pattern, weather prediction etc., for the next day, it is always liable to have significant deviations from the given schedule.

These features being inherent need to be accommodated for fulfilling the stated goal of maximising the electricity generation from renewable energy sources. Still, because of these features, (actually drawbacks compared to other types of generation) the utilities have a hesitation in absorbing (and paying for) the renewable generation, while Government and regulatory authorities want the utilities to absorb more and more renewable energy, leading to a conflict.

The first issue is the pricing of renewable energy, and there are two distinct ways to go about it. One is that utilities be required to pay a rate to the generator which covers the latter's cost, and the price differential between the renewable energy and conventional energy is covered either by Government subsidy or by uplift in consumer tariff. In this approach, rate fixation is very subjective on many counts, and regulators may find it difficult to specify a rate which reasonably covers the generation cost and does not impose an unreasonable burden on the Government, utilities or the consumer.

The other way of pricing renewable energy is to pay for it at the system marginal price. This, as compared to the first approach has many advantages, and it is very easily possible to adopt it in India, through the established mechanism of UI mentioned earlier. In this scheme, the entire injection of renewable energy could be treated as UI, and paid for at the prevailing UI rate. Alternatively, the scheduled injection could be paid for at a regulator specified or contracted rate, and the deviation from schedule treated as UI.

The other issue pertains to scheduling. In western countries, scheduling of renewable energy, or rather the inevitable deviations from schedule pose two major problems in grid operation. One is that of commercial treatment of the deviations, which cannot be neutralised in kind. This is not a problem in India at all. All deviations can be easily accounted as UI in an objective and dispute-free manner. The other problem is that RE's deviation from schedule (or variability of injection) give rise to load-generation imbalance in the control area in which they are embedded, and cause fluctuations of control area's net interchange. In countries having substantial wind generation, this constitutes a serious problem, causing power flow variations throughout the inter-connection, as also frequency fluctuations. Since India already has a full-fledged mechanism for handling deviations from interchange schedules and also allows frequency to fluctuate, the above is just not a problem in India. It is therefore rather unfortunate that under the influence of western philosophies and practices, measures are being proposed to curb or neutralise the inevitable deviations of renewable energy injection. One can only hope that unnecessary costs and complications would not be created.

Another beauty of UI mechanism, not fully appreciated, is that it works back-to-back. For example, if the injection of a generating station into its host control area is 10 MW short of schedule, and the control area consequently overdraws 10 MW from the larger grid, whatever UI charges the control area has to pay to the larger grid, it recovers the same from the defaulting generating station.

Figure 3.1

Frequency Profile in Northern Region for: 05-04-2011

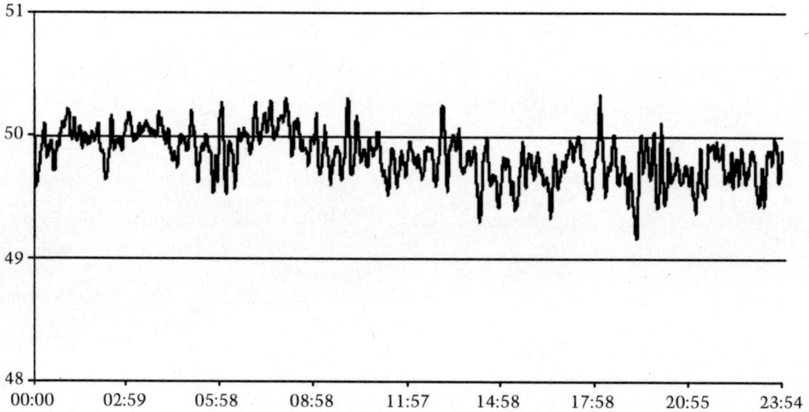

Note: The figure shown above are indicative only and are calculated based on data acquired from remote locations in Northern Region over existing communication links.

Figure 3.2

Frequency Profile in Northern Region for: 22-04-2011

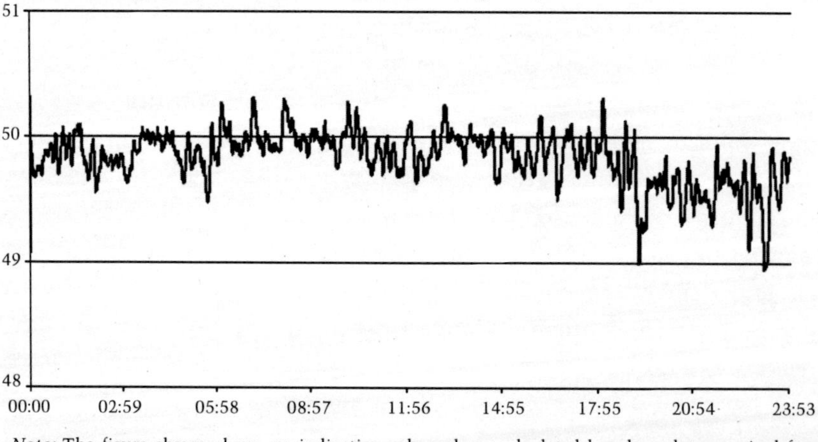

Note: The figure shown above are indicative only and are calculated based on data acquired from remote locations in Northern Region over existing communication links.

Annexure A-3.1

Extracts from the Indian Electricity Grid Code (Notified by CERC on 28.04.2010)

6.4 Demarcation of Responsibilities

1. The national interconnected grid is divided into control areas, like Regional ISTS, States, DVC, etc. where the load despatch centre or system operator of the respective control area controls its generation and/or load to maintain its interchange schedule with other control areas whenever required to do so and contributes to frequency regulation of the synchronously operating system. The Load Despatch Centre of a control area therefore is responsible for coordinating the scheduling of a generating station, within the control area, real-time monitoring of the station's operation, checking that there is no gaming (gaming is an intentional mis-declaration of a parameter related to commercial mechanism in vogue, in order to make an undue commercial gain) in its availability declaration, or in any other way revision of availability declaration and injection schedule, switching instructions, metering and energy accounting, issuance of UI accounts within the control area, collections/disbursement of UI payments, outage planning, etc. The following clause gives the criteria for demarcation of control area jurisdiction.

2. The following generating stations shall come under the respective Regional ISTS control area and hence the respective RLDC shall coordinate the scheduling of the following generating stations :

a) Central Generating Stations (excluding stations where full Share is allocated to host state),

b) Ultra-Mega power projects,

c) In other cases, the control area shall be decided on the following criteria:

(i) If a generating station is connected only to the ISTS, RLDC shall coordinate the scheduling, except for Central Generating Stations where full Share is allocated to one State.

(ii) If a generating station is connected only to the State transmission network, the SLDC shall coordinate scheduling, except for the case as at (a) above.

(iii) If a generating station is connected both to ISTS and the State network, scheduling and other functions performed by the system operator of a control area will be done by SLDC only. If state has more than 50 per cent Share of power, the role of concerned RLDC, in such a case, shall be limited to consideration of the schedule for inter-state exchange of power on account of this ISGS while determining the net drawal schedules of the respective states. If the State has a Share of 50 per cent or less, the scheduling and other functions shall be performed by RLDC.

(iv) In case commissioning of a plant is done in stages the decision regarding scheduling and other functions performed by the system operator of a control area would be taken on the basis of above criteria depending on generating capacity put into commercial operation at that point of time. Therefore it could happen that the plant may be in one control area (i.e., SLDC) at one point of time and another control area (i.e., RLDC) at another point of time. The switch over of control area would be done expeditiously after the change, w.e.f. the next billing period.

3. There may be exceptions with respect to above provisions, for reasons of operational expediency, subject to approval of CERC. Irrespective of the control area the jurisdiction, if a generating station is connected both to the ISTS and the STU, the load despatch centre of the control area under whose jurisdiction the generating station falls, shall take into account grid security implication in the control area of the other load despatch centre

4. For those generating station supplying power to any state other than host state and whose scheduling is not coordinated by RLDC, the role of the concerned RLDC shall be limited to consideration of the schedule for inter-State exchange of power on account of this generating station while determining the net drawal schedules of the respective control area.

5. The Regional grids shall be operated as power pools with decentralised scheduling and despatch, in which the States shall have operational autonomy, and SLDCs shall have the total responsibility for

(i) scheduling/despatching their own generation (including generation of their embedded licensees),

(ii) regulating the demand of its control area,

(iii) scheduling their drawal from the ISGS (within their share in the respective plant's expected capability),

(iv) permitting long-term access, medium-term and short-term open access transactions for embedded generators/consumers, in accordance with the contracts, and

(v) regulating the net drawal of their control area from the regional grid in accordance with the respective regulations of the CERC.

6. The system of each regional entity shall be treated and operated as a notional control area. The algebraic summation of scheduled drawal from ISGS and from contracts through a long-term access, medium-term and short-term open access arrangements shall provide the drawal schedule of each regional entity, and this shall be determined in advance on day-ahead basis. The regional entities shall regulate their generation and/or consumers' load so as to maintain their actual drawal

from the regional grid close to the above schedule. If regional entities deviate from the drawal schedule, within the limit specified by the CERC in UI Regulations as long as such deviations do not cause system parameters to deteriorate beyond permissible limits and/or do not lead to unacceptable line loading, However,such deviations from net drawal schedule shall be priced through the Unscheduled Interchange (UI) mechanism.

7. The SLDC, SEB/distribution licensee shall always endeavour to restrict the net drawal of the state from the grid to within the drawal schedules, whenever the system frequency is below 49.7 Hz. The concerned SEB/distribution licensee User,SLDC shall ensure that their automatic demand management scheme mentioned in clause 5.4.2 acts to ensure that there is no over-drawl when frequency is 49.5 Hz or below. If the automatic demand management scheme has not yet been commissioned, then action has to be taken as per manual demand management scheme to ensure zero overdrawal when frequency r is 49.5 Hz. or below.

8. The SLDCs/STUs/Distrbution Licensees shall regularly carry out the necessary exercises regarding short-term demand estimation for their respective States/area, to enable them to plan in advance as to how they would meet their consumers' load without overdrawing from the grid.

9. The ISGS , other generating stations and sellers shall be responsible for power generation/power injection generally according to the daily schedules advised to them by the RLDC/SLDC on the basis of the contracts/requisitions received from the SLDCs/ buyers/Power Exchanges.

10. The ISGS would normally be expected to generate power according to the daily schedules advised to them. The ISGS may also deviate from the given schedules within the limits specified in the UI Regulations of CERC, depending on the plant and system conditions. In particular, they may be allowed to generate beyond the given schedule under deficit conditions as long as such deviations do not cause system parameters to deteriorate beyond permissible limits and/or do not lead to unacceptable line loading. Deviations, if any, from the ex-power plant generation schedules shall be appropriately priced in accordance with UI Regulations. In addition, deviations, from schedules causing congestion, shall also be priced in accordance with the Congestion Charge Regulations of CERC.

11. Provided that when the frequency is higher than 50.2 Hz, the actual net injection shall not exceed the scheduled despatch for that time block. Also, while the frequency is above 50.2 Hz, the ISGS may (at their discretion) back down without waiting for an advice from RLDC to restrict the frequency rise. When the frequency falls below 49.7 Hz, the generation at all ISGS (except those on peaking duty) shall be maximised, at least upto the level to which can be sustained, without waiting for an advice from RLDC subject to the condition that such increase does not lead to unacceptable line loading or system parameters to deteriorate beyond permissible limit.

12. However, notwithstanding the above, the RLDC may direct the SLDCs/ISGS/ other regional entities to increase/decrease their drawal/generation in case of contingencies e.g., overloading of lines/transformers, abnormal voltages, threat to system security. Such directions shall immediately be acted upon. In case the situation does not call for very urgent action, and RLDC has some time for analysis, it shall be checked whether the situation has arisen due to deviations from schedules, pursuant to short-term open access. These shall be got terminated first, before an action, which would affect the scheduled supplies to the long term and medium-term customers is initiated in accordance with Central Electricity Regulatory Commission (Grant of Connectivity, Long-term Access and Medium-term Open Access in Inter-state Transmission and related matters) Regulations, 2009.

13. For all outages of generation and transmission system, which may have an effect on the regional grid, all Regional entities shall cooperate with each other and coordinate their actions through Operational Coordination Committee (OCC) for outages foreseen sufficiently in advance and through RLDC (in all other cases), as per procedures finalised separately by OCC. In particular, outages requiring restriction of ISGS generation and/or restriction of ISGS Share which a beneficiary can receive and curtailment of other long-term transactions shall be planned carefully to achieve the best optimisation.

14. The regional entities shall enter into separate joint/bilateral agreement(s) to identify the beneficiary's Shares in ISGS (based on the allocations by the Govt. of India, where applicable), scheduled drawal pattern, tariffs, payment terms etc. All such agreements shall be filed with the concerned RLDC(s) and RPC, Secretariat, for being considered in scheduling and regional energy accounting. Any bilateral agreements between buyer and seller for scheduled interchanges on long-term, medium–term basis shall also specify the interchange schedule, which shall be duly filed with CTU and CTU shall inform RLDC and SLDC, as the case may be about these agreements in accordance with Central Electricity Regulatory Commission (Grant of Connectivity, Long-term Access and Medium-term Open Access in inter-state Transmission and related matters) Regulations, 2009.

15. All other regional entities should abide by the concept of frequency-linked load despatch and pricing of deviations from schedule, i.e., unscheduled interchanges. All regional entities should normally be operated according to the standing frequency-linked load despatch guidelines issued by the RLDC, to the extent possible, unless otherwise advised by the RLDC/SLDC.

16. The ISGS shall make an advance declaration of ex-power plant MW and MWh capabilities foreseen for the next day, i.e., from 0000 hrs to 2400 hrs. During fuel shortage condition, in case of thermal stations, they may specify minimum MW, maximum MW, MWh capability and declaration of fuel shortage. The generating

stations shall also declare the possible ramping up/ramping down in a block. In case of a gas turbine generating station or a combined cycle generating station, the generating station shall declare the capacity for units and modules on APM gas, RLNG and liquid fuel separately, and these shall be scheduled separately.

17. While making or revising its declaration of capability, except in case of Run Off the River (with up to three hour pondage) hydro stations, the ISGS shall ensure that the declared capability during peak hours is not less than that during other hours. However, exception to this rule shall be allowed in case of tripping/re-synchronisation of units as a result of forced outage of units.

18. It shall be incumbent upon the ISGS to declare the plant capabilities faithfully, i.e., according to their best assessment. In case, it is suspected that they have deliberately over/under declared the plant capability contemplating to deviate from the schedules given on the basis of their capability declarations (and thus make money either as undue capacity charge or as the charge for deviations from schedule), the RLDC may ask the ISGS to explain the situation with necessary backup data.

19. The ISGS shall be required to demonstrate the declared capability of its generating station as and when asked by the Regional Load Despatch Centre of the region in which the ISGS is situated. In the event of the ISGS failing to demonstrate the declared capability, the capacity charges due to the generator shall be reduced as a measure of penalty.

20. The quantum of penalty for the first mis-declaration for any duration/ block in a day shall be the charges corresponding to two days fixed charges. For the second mis-declaration the penalty shall be equivalent to fixed charges for four days and for subsequent mis-declarations, the penalty shall be multiplied in the geometrical progression over a period of a month.

21. The CTU shall install special energy meters on all inter-connections between the regional entities and other identified points for recording of actual net MWh interchanges and MVArh drawals. The installation, operation and maintenance of special energy meters shall be in accordance with Central Electricity Authority (Installation and Operation of Meters) Regulations, 2006. All concerned entities (in whose premises the special energy meters are installed) shall take weekly meter readings and transmit them to the RLDC by Tuesday noon. The SLDC must ensure that the meter data from all installations within their control area are transmitted to the RLDC within the above schedule.

22. The RLDC shall be responsible for computation of actual net injection/ drawal of concerned regional entities, 15 minute-wise, based on the above meter readings. The above data along with the processed data of meters shall be forwarded by the RLDC to the RPC secretariat on a weekly basis by each Thursday noon for the seven day period ending on the previous Sunday mid-night, to enable the latter to

prepare and issue the Unscheduled inter-change (UI) account in accordance with the CERC (Unschesduled Interchange charges and related matters) Regulations, 2010, as amended from time to time. All computations carried out by RLDC shall be open to all regional entities for checking/verifications for a period of 15 days. In case any mistake/ omission is detected, the RLDC shall forthwith make a complete check and rectify the same.

23. The operating log books of the generating station shall be available for review by the Regional Power Committee. These books shall keep record of machine operation and maintenance.

24. Hydro generating stations are expected to respond to grid frequency changes and inflow fluctuations. The hydro generating stations shall be free to deviate from the given schedule without causing grid constraint and a compensation for difference between the actual net energy supply by the hydro generating station and the scheduled energy (ex-bus) over day shall be made by the concerned Regional Load Despatch Centre in the day ahead schedule for the 4th day (day plus 3).

25. RLDC shall periodically review the actual deviation from the despatch and net drawal schedules being issued, to check whether any of the regional entities are indulging in unfair gaming or collusion. In case any such practice is detected, the matter shall be reported to the Member Secretary, RPC for further investigation/ action.

26. NLDC shall be responsible for scheduling and despatch of electricity over inter-regional links in accordance with the grid code specified by Central Commission in coordination with Regional Load Despatch Centers. NLDC shall be responsible for coordination with Regional Load Despatch Centers for the energy accounting of inter-regional exchange of power. NLDC shall also be responsible for coordination for trans-national exchange of power.

27. NLDC shall develop a procedure for scheduling of collective transaction through Power Exchanges, scheduling of inter-regional power exchanges including HVDC setting responsibility and power exchanges of the country with other countries.

6.5 Scheduling and Despatch Procedure for Long-Term Access, Medium-Term and Short-Term Open Access

(To be read with provisions of Open Access Regulations 2008 as amended from time to time. The scheduling procedure for medium-term open access transactions shall be similar to the scheduling procedure for long-term access transactions and is as given below, except where it is specifically mentioned for collective transactions):

1. All inter-state generating stations (ISGS) shall be duly listed on the respective RLDC and SLDC web-sites. The station capacities and allocated/contracted Shares of different beneficiaries shall also be listed out.

2. Each State shall be entitled to a MW despatch up to (foreseen expower plant MW capability for the day) x (State's Share in the station's capacity) for all such stations. In case of hydro-electric stations, there would also be a limit on daily MWh despatch equal to (MWh generation capacity for the day)X (State's Share in the station's capacity).

3. By 8 AM every day, the ISGS shall advise the concerned RLDC, the station-wise ex-power plant MW and MWh capabilities foreseen for the next day, i.e., from 0000 hrs to 2400 hrs of the following day.

4. The above information of the foreseen capabilities of the ISGS and the corresponding MW and MWh entitlements of each State, shall be compiled by the RLDC every day for the next day, and advised to all beneficiaries by 10 AM. The SLDCs shall review it *vis-à-vis* their foreseen load pattern and their own generating capability including bilateral exchanges, if any, and advise the RLDC by 3 PM their drawal schedule for each of the ISGS in which they have Shares, long-term and medium-term bilateral interchanges, approved short-term bilateral interchanges.

5. Scheduling of collective transaction:

a. NLDC shall indicate to Power Exchange(s), the list of interfaces/control areas/regional transmission systems on which unconstrained flows are required to be advised by the Power Exchange(s) to the NLDC. Power Exchange(s) shall furnish the interchange on various interfaces/control areas/regional transmission systems as intimated by NLDC. Power Exchange(s) shall also furnish the information of total drawal and injection in each of the regions. Based on the information furnished by the Power Exchanges, NLDC shall check for congestion. In case of congestion, NLDC shall inform the Exchanges about the period of congestion and the available limit for scheduling of collective transaction on respective interface/control area/transmission systems during the period of congestion for Scheduling of Collective Transaction through the respective Power Exchange. The limit for scheduling of collective transaction for respective Power Exchange shall be worked out in accordance with CERC directives. Based on the application for scheduling of Collective Transaction submitted by the Power Exchange(s), NLDC shall send the details (Scheduling Request of Collective Transaction) to different RLDCs for final checking and incorporating them in their schedules. After getting confirmation from RLDCs, NLDC shall convey the acceptance of scheduling of collective transaction to Power Exchange(s). RLDCs shall schedule the Collective Transaction at the respective periphery of the Regional Entities.

b. The individual transactions for State Utilities/intra-State Entities shall be scheduled by the respective SLDCs. Power Exchange(s) shall send the detailed break up of each point of injection and each point of drawal within the State to respective SLDCs after receipt of acceptance from NLDC. Power Exchange(s) shall ensure necessary coordination with SLDCs for scheduling of the transactions.

c. Timeline for above activities will be as per detailed procedure for Scheduling of Collective Transaction issued in accordance with CERC (Open access in inter-state transmission) Regulations, 2008 and as amended from time to time.

6. The SLDCs may also give standing instructions to the RLDC such that the RLDC itself may decide the best drawal schedules for the States.

7. By 6 PM each day, the RLDC shall convey:

(i) The ex-power plant 'despatch schedule' to each of the ISGS, in MW for different time block, for the next day. The summation of the ex-power plant drawal schedules advised by all beneficiaries shall constitute the ex-power plant station-wise despatch schedule.

(ii) The 'net drawal schedule' to each regional entity, in MW for different time block, for the next day. The summation of the station-wise ex-power plant drawal schedules from all ISGS and drawal from/injection to regional grid consequent to other long-term access, medium-term and short-term open access transactions, after deducting the transmission losses (estimated), shall constitute the regional entity-wise drawal schedule.

8. The SLDCs/ISGS shall inform any modifications/changes to be made in drawal schedule/foreseen capabilities, if any, to RLDC by 10 PM or preferably earlier.

9. The hydro electric generation stations are expected to respond to grid frequency changes and inflow fluctuations. They would, therefore, be free to deviate from the given schedule as long as they do not cause a grid constraint. As a result, the actual net energy supply by a hydro generating station over a day may differ from schedule energy (ex-bus) for that day. A compensation shall then be made by the concerned load despatch centre in the day ahead schedule for the 4th day (day plus 3).

10. The declaration of the generating capability by hydro ISGS should include limitation on generation during specific time periods, if any, on account of restriction(s) on water use due to irrigation, drinking water, industrial, environmental considerations etc. The concerned Load Despatch Centre shall periodically check that the generating station is declaring the capacity and energy sincerely, and is not manipulating the declaration with the intent of making undue money through UI.

11. Since variation of generation in run-of-river power stations shall lead to spillage, these shall be treated as must run stations. All renewable energy power plants, except for biomass power plants, and non-fossil fuel based cogeneration plants whose

tariff is determined by the CERC shall be treated as 'MUST RUN' power plants and shall not be subjected to 'merit order despatch' principles.

12. Run-of-river power station with pondage and storage type power stations are designed to operate during peak hours to meet system peak demand. Maximum capacity of the station declared for the day shall be equal to the installed capacity including overload capability, if any, minus auxiliary consumption, corrected for the reservoir level. The Regional Load Despatch Centers shall ensure that generation schedules of such type of stations are prepared and the stations despatched for optimum utilisation of available hydro energy except in the event of specific system requirements/constraints.

13. The schedule finalised by the concerned load despatch centre for hydro generating station, shall normally be such that the scheduled energy for a day equals the total energy (ex-bus) expected to be available on that day, as declared by the generating station, based on foreseen/planned water availability/release. It is also expected that the total net energy actually supplied by the generating station on that day would equal the declared total energy, in order that the water release requirement is met. While the 15-minute wise, deviations from schedule would be accounted for as Unscheduled Interchange (UI), the net energy deviation for the whole day, if any, shall be additionally accounted for as shown in the illustration.

Illustration

Suppose the foreseen/expected total energy (ex-bus) for Day-1 is E1, the scheduled energy is S1, actual net energy (metered) is A1, all in ex-bus MWh. Suppose the expected energy availability for Day 4, as declared by the generator, is E4. Then, the schedule for day 4 shall be drawn up such that the scheduled energy for Day 4, shall be

S4=E4+(A1-E1), Similarly,

S5= E5+(A2-E2),

S6=E6+(A3-E3),

S7=E7+(A4-E4), and so on.

14. While finalising the above daily despatch schedules for the ISGS, RLDC shall ensure that the same are operationally reasonable, particularly in terms of ramping-up/ramping-down rates and the ratio between minimum and maximum generation levels. A ramping rate of upto 200 MW per hour should generally be acceptable for an ISGS and for a regional entity (50 MW in NER), except for hydro-electric generating stations which may be able to ramp up/ramp down at a faster rate.

15. While finalising the drawal and despatch schedules as above, the RLDC shall also check that the resulting power flows do not give rise to any transmission

constraints. In case any impermissible constraints are foreseen, the RLDC shall moderate the schedules to the required extent, under intimation to the concerned regional entities. Any changes in the scheduled quantum of power which are too fast or involve unacceptably large steps, may be converted into suitable ramps by the RLDC.

16. In the event of bottleneck in evacuation of power due to any constraint, outage, failure or limitation in the transmission system, associated switchyard and substations owned by the Central Transmission Utility or any other transmission licensee involved in inter-state transmission (as certified by the RLDC) necessitating reduction in generation, the RLDC shall revise the schedules which shall become effective from the 4th time block, counting the time block in which the bottleneck in evacuation of power has taken place to be the first one. Also, during the first, second and third time blocks of such an event, the scheduled generation of the ISGS shall be deemed to have been revised to be equal to actual generation, and the scheduled drawals of the beneficiaries shall be deemed to have been revised accordingly.

17. In case of any grid disturbance, scheduled generation of all the ISGS and scheduled drawal of all the beneficiaries shall be deemed to have been revised to be equal to their actual generation/drawal for all the time blocks affected by the grid disturbance. Certification of grid disturbance and its duration shall be done by the RLDC.

18. Revision of declared capability by the ISGS(s) having two part tariff with capacity charge and energy charge(except hydro stations) and requisition by beneficiary(ies) for the remaining period of the day shall also be permitted with advance notice. Revised schedules/declared capability in such cases shall become effective from the 6th time block, counting the time block in which the request for revision has been received in the RLDC to be the first one. Provided that RLDC may allow revision, of the DC at 6 hourly intervals effective form 0000,0600,1200 and 1800 hours in case of Run of the River (ROR) and pondage based hydro generating stations, if there is large variation of expected energy (MWh) for the day compared to previous declaration.

19. Notwithstanding anything contained in Regulation 6.5(18), in case of forced outage of a unit for a Short-Term Bilateral Transaction, where a generator of capacity of 100 MW and above is seller, the generator shall immediately intimate the same along with the requisition for revision of schedule and estimated time of restoration of the unit, to SLDC/RLDC as the case may be. With the objective of not affecting the existing contracts, the revision of schedule shall be with the consent of the buyer till 31.07.2010. Thereafter, consent of the buyer shall not be a pre-requisite for such revision of schedule. The schedule of the generator and the buyer shall be revised, accordingly. The revised schedules shall become effective from the 4th time block, counting the time block in which the forced outage is declared to be

the first one. The RLDC shall inform the revised schedule to the seller and the buyer. The original schedule shall become effective from the estimated time of restoration of the unit. However the transmission charges as per original schedule shall continue to be paid for two days.

20. If, at any point of time, the RLDC observes that there is need for revision of the schedules in the interest of better system operation, it may do so on its own, and in such cases, the revised schedules shall become effective from the 4th time block, counting the time block in which the revised schedule is issued by the RLDC to be the first one.

21. To discourage frivolous revisions, an RLDC may, at its sole discretion, refuse to accept schedule/capability changes of less than two (2) percent of previous schedule/capability. The schedule of thermal generating stations indicating fuel shortage while intimating the Declared Capacity to the RLDC shall not be revised except in case of forced outage of generating unit. Provided that in case of gas based ISGS, for optimum utilisation of gas, this shall be permitted, i.e., in case of tripping of a unit, this gas may be diverted to another unit using the same gas.

22. The Regional Load Despatch Centre shall also formulate the procedure for meeting contingencies both in the long run and in the short run (Daily scheduling).

23. Special dispensation for scheduling of wind and solar generation.

(i) With effect from 1.1.2011 Scheduling of wind power generation plants would have to be done for the purpose of UI where the sum of generation capacity of such plants connected at the connection point to the transmission or distribution system is 10 MW and above and connection point is 33 KV and above, and where PPA has not yet been signed. For capacity and voltage level below this, as well as for old wind farms (A wind farm is collection of wind turbine generators that are connected to a common connection point) it could be mutually decided between the Wind Generator and the transmission or distribution utility, as the case may be, if there is no existing contractual agreement to the contrary. The schedule by wind power generating stations may be revised by giving advance notice to SLDC/RLDC, as the case may be. Such revisions by wind power generating stations shall be effective from 6th time-block, the first being the time–block in which notice was given. There may be maximum of 8 revisions for each 3 hour time slot starting from 00:00 hours during the day.

(ii) The schedule of solar generation shall be given by the generator based on availability of the generator, weather forecasting, solar insolation, season and normal solar generation curve and shall be vetted by the RLDC in which the generator is located and incorporated in the inter-state schedule...

If RLDC is of the opinion that the schedule is not realistic , it may ask the solar generator to modify the schedule.

(iii) Concerned RLDC and SLDC shall maintain the record of schedule from renewable power generating stations based on type of renewable energy sources i.e., wind or solar from the point of view of grid security. While scheduling generating stations in a region, system operator shall aim at utilising available wind and solar energy fully.

24. Generation schedules and drawal schedules issued/revised by the Regional Load Despatch Centre shall become effective from designated time-block irrespective of communication success.

25. For any revision of scheduled generation, including *post facto* deemed revision, there shall be a corresponding revision of scheduled drawals of the beneficiaries.

26. A procedure for recording the communication regarding changes to schedules duly taking into account the time factor shall be evolved by the Central Transmission Utility.

27. When for the reason of transmission constraints e.g. congestion or in the interest of grid security, it becomes necessary to curtail power flow on a transmission corridor, the transactions already scheduled may be curtailed by the Regional Load Despatch Centre.

28. The short-term customer shall be curtailed first followed by the medium-term customers, which shall be followed by the long-term customers and amongst the customers of a particular category, curtailment shall be carried out on *pro rata* basis.

29. After the operating day is over at 2400 hours, the schedule finally implemented during the day (taking into account all before-the-fact changes in despatch schedule of generating stations and drawal schedule of the States) shall be issued by RLDC. These schedules shall be the datum for commercial accounting. The average ex-bus capability for each ISGS shall also be worked out based on all before-the-fact advice to RLDC.

30. Collective Transaction through Power Exchange(s) would normally be curtailed subsequent to the Short Term Bilateral Transaction(s).

31. RLDCs would curtail a Transaction at the periphery of the Regional Entities. SLDC(s) shall further incorporate the inter-se curtailment of intra-State Entities to implement the curtailment.

32. RLDC shall properly document all above information i.e., station-wise foreseen ex-power plant capabilities advised by the generating stations, the drawal

schedules advised by regional entities, all schedules issued by the RLDC, and all revisions/updating of the above.

33. The procedure for scheduling and the final schedules issued by RLDC, shall be open to all regional entities and other regional open access customers entities for any checking/verification, for a period of 5 days. In case any mistake/omission is detected, the RLDC shall forthwith make a complete check and rectify the same.

34. While availability declaration by ISGS shall have a resolution of one (1) MW and one (1) MWh, all entitlements, requisitions and schedules shall be rounded off to the nearest two decimal at each control area boundary for each of the transaction, to have a resolution of 0.01 MW and 0.01 MWh.

6.6 Reactive Power and Voltage Control

1. Reactive power compensation should ideally be provided locally, by generating reactive power as close to the reactive power consumption as possible. The Regional Entities except Generating Stations are therefore expected to provide local VAr compensation/generation such that they do not draw VArs from the EHV grid, particularly under low-voltage condition. To discourage VAr drawals by Regional Entities except Generating Stations, VAr exchanges with ISTS shall be priced as follows:

- The Regional Entity except Generating Stations pays for VAr drawal when voltage at the metering point is below 97 per cent.
- The Regional Entity except Generating Stations gets paid for VAr return when voltage is below 97 per cent.
- The Regional Entity except Generating Stations gets paid for VAr drawal when voltage is above103 per cent.
- The Regional Entity except Generating Stations pays for VAr return when voltage is above 103 per cent.
- Provided that there shall be no charge/payment for VAr drawal/return by a Regional Entity except Generating Stations on its own line emanating directly from an ISGS.

2. The charge for VArh shall be at the rate of 10 paise/kVArh w.e.f. 1.4.2010, and this will be applicable between the Regional Entity, except Generating Stations, and the regional pool account for VAr interchanges. This rate shall be escalated at 0.5paise/kVArh per year thereafter, unless otherwise revised by the Commission.

3. Notwithstanding the above, RLDC may direct a Regional Entity except Generating Stations to curtail its VAr drawal/injection in case the security of grid or safety of any equipment is endangered.

4. In general, the Regional Entities except Generating Stations shall endeavor to minimise the VAr drawal at an interchange point when the voltage at that point is below 95 per cent of rated, and shall not return VAr when the voltage is above 105 per

cent. ICT taps at the respective drawal points may be changed to control the VAr interchange as per a Regional Entity except Generating Stations's request to the RLDC, but only at reasonable intervals.

5. Switching in/out of all 400 kV bus and line Reactors throughout the grid shall be carried out as per instructions of RLDC. Tap changing on all 400/220 kV ICTs shall also be done as per RLDCs instructions only.

6. The ISGS and other generating stations connected to regional grid shall generate/absorb reactive power as per instructions of RLDC, within capability limits of the respective generating units, that is without sacrificing on the active generation required at that time. No payments shall be made to the generating companies for such VAr generation/absorption.

7. VAr exchange directly between two Regional Entities except Generating Stations on the interconnecting lines owned by them (singly or jointly) generally address or cause a local voltage problem, and generally do not have an impact on the voltage profile of the regional grid. Accordingly, the management/control and commercial handling of the VAr exchanges on such lines shall be as per following provisions, on case-by-case basis:

i) The two concerned Regional Entities except Generating Stations may mutually agree not to have any charge/payment for VAr exchanges between them on an interconnecting line.

ii) The two concerned Regional Entities except Generating Stations may mutually agree to adopt a payment rate/scheme for VAr exchanges between them identical to or at variance from that specified by CERC for VAr exchanges with ISTS. If the agreed scheme requires any additional metering, the same shall be arranged by the concerned Beneficiaries.

iii) In case of a disagreement between the concerned Regional Entities except Generating Stations (e.g. one party wanting to have the charge/payment for VAr exchanges, and the other party refusing to have the scheme), the scheme as specified in Annexure A-3.2 shall be applied. The per kVArh rate shall be as specified by CERC for VAr exchanges with ISTS

iv) The computation and payments for such VAr exchanges

Annexure to IEGC (Referred to in Section 6.1 (d))

Complementary Commercial Mechanisms

1. The beneficiaries shall pay to the respective ISGS Capacity charges corresponding to plant availability and/or Energy charges for the scheduled despatch, in accordance with the relevant contracts/orders of CERC. The bills for these charges shall be issued by the respective ISGS to each beneficiary on monthly basis.

2. The sum of the above two charges from all beneficiaries shall fully reimburse the ISGS for generation according to the given despatch schedule. In case of a deviation in actual generation from the despatch schedule, the concerned ISGS shall receive or shall pay in accordance with UI regulation of CERC. Similarly, the deviation of actual drawl by any regional entity from the net drawl schedule shall be treated as UI. All 15-minute energy figures (net scheduled, actually metered and UI) shall be rounded off to the nearest 0.01 MWh. The UI charges and the modalities of settlement of UI shall be in accordance with UI Regulation of CERC.

3. Wind energy being of variable nature, needs to be predicted with reasonable accuracy for proper scheduling and dispatching of power from these sources in the interconnected system. Hence wind generation forecasting is necessary for increased penetration. Wind generation forecasting can be done on an individual developer basis or joint basis for an aggregated generation capacity of 10 MW and connected at a connection point of 33 kV and above. If done jointly, the wind forecasting facility shall be built and operated by wind developers in the area and sharing of the cost shall be mutually discussed and agreed.

4. The wind energy forecasting system shall forecast power based on wind flow data at the following time intervals:

i) Day ahead forecast: Wind/power forecast with an interval of 15 minutes for the next 24 hours for the aggregate generation capacity of 10 MW and above.

ii) The schedule by such wind power generating stations supplying interstate power under long-term access and medium-term and short-term open access may be revised by giving advance notice to RLDC. Such revisions by wind power generating stations shall be effective from 6th time-block, the first being the time-block in which notice was given.

There may be maximum of 8 revisions for each 3-hour time slot starting from 00:00 hours during the day.

5. The wind generators shall be responsible for forecasting their generation upto an accuracy of 70 per cent. Therefore, if the actual generation is beyond +/- 30 per cent of the schedule, wind generator would have to bear the UI charges. For actual generation within +/- 30 per cent of the schedule, no UI would be payable/receivable by Generator. The host State shall bear the UI charges for this variation, i.e., within +/- 30 per cent. However, the UI charges borne by the host State due to the wind generation, shall be shared among all the States of the country in the ratio of their peak demands in the previous month based on the data published by CEA, in the form of a regulatory charge known as the Renewable Regulatory Charge operated through the Renewable Regulatory Fund (RRF). This provision shall be applicable with effect from 1.1.2011, for new wind farms with collective capacity of 10 MW and above connected at connection point of 33 KV level and above, and who have

not signed any PPA with states or others as on the date of coming into force of this IEGC. Illustrative calculations in respect of above mechanism are given in Annexure.

6. A maximum generation of 150 per cent of the schedule only, would be allowed in a time-block, for injection by wind, from the grid security point of view. For any generation above 150 per cent of schedule, if grid security is not affected by the generation above 150 per cent, the only charge payable to the wind energy generator would be the UI charge applicable corresponding to 50- 50.02 HZ .

7. In case of solar generation no UI shall be payable/receivable by Generator. The host State, shall bear the UI charges for any deviation in actual generation from the schedule. However, the net UI charges borne by the host State due to the solar generation, shall be shared among all the States of the country in the ratio of their peak demands in the previous month based on the data published by CEA, in the form of regulatory charge known as the Renewable Regulatory Charge operated through the Renewable Regulatory Fund as referred to in clause 5 above. This provision shall be applicable, with effect from 1.1.2011, for new solar generating plants with capacity of 5 MW and above connected at connection point of 33 KV level and above and, who have not signed any PPA with states or others as on the date of coming into force of this IEGC. Illustrative calculations in respect of above mechanism are given in Annexure.

8. All Regional Energy Accounting calculations carried out by RPC Secretariats shall be open to all regional entities for any checking/verification for a period of 15 days. In case any mistake is detected, RPC Secretariats shall forthwith make a complete check and rectify the mistakes.

9. NLDC shall prepare, within one month of notification of these regulations, a detailed procedure for implementation of the mechanism of Renewable Regulatory Fund and submit the same for approval by the Commission.

10. Regional Energy Accounts on monthly basis shall be prepared and issued by the RPC Secretariats for the purpose of billing and payment of various charges. Regional Energy Account for a month shall be issued in the following month based on the data provided by RLDC.

11. RPC Secretariats shall also issue the weekly statement for VAR charges, to all regional entities who have a net drawal/injection of reactive energy under low/high voltage conditions. These payments shall also have a high priority and the concerned regional entities and other regional entities shall pay the indicated amounts into regional reactive pool account operated by the RLDC within 10 (ten) days of statement issue, provided that the Commission may direct any entity other than RLDC to operate the regional reactive pool account. The regional entities who have to receive the money on account of VAR charges would then be paid out from the regional reactive pool account, within two (2) working days from the receipt of payment in the Reactive pool account.

12. If payments against the above VAr charges are delayed by more than two days, i.e., beyond twelve (12) days from statement issue, the defaulting regional entity shall have to pay simple interest @ 0.04 per cent for each day of delay. The interest so collected shall be paid to the regional entities who had to receive the amount, payment of which got delayed. Persistent payment defaults, if any, shall be reported by the RLDC to the Member Secretary, RPC, for initiating remedial action.

13. The money remaining in the regional reactive account after payout of all VAr charges upto 31st March of every year shall be utilised for training of the SLDC operators, and other similar purposes which would help in improving/streamlining the operation of the respective regional grids, as decided by the respective RPC from time to time.

14. In case the voltage profile of a regional grid improves to an extent that the total pay-out from the regional VAr charges account for a week exceeds the total amount being paid-in for that week, and if the regional reactive account has no balance to meet the deficit, the pay-outs shall be proportionately reduced according to the total money available in the above account.

15. The RLDC shall table the complete statement of the regional UI account and the regional Reactive Energy pool account in the RPC's Commercial Committee meeting, on a quarterly basis, for audit by the latter.

Annexure A-3.2

What is Availability Based Tariff?

The term Availability Based Tariff, particularly in the Indian context, stands for a rational tariff structure for power supply from generating stations, on a contracted basis. The power plants have fixed and variable costs. The fixed cost elements are interest on loan, return on equity, depreciation, O&M expenses, insurance, taxes and interest on working capital. The variable cost comprises of the fuel cost, i.e., coal and oil in case of thermal plants and nuclear fuel in case of nuclear plants. In the Availability Based Tariff mechanism, the fixed and variable cost components are treated separately. The payment of fixed cost to the generating company is linked to availability of the plant, that is, its capability to deliver MWs on a day-by-day basis. The total amount payable to the generating company over a year towards the fixed cost depends on the average availability (MW delivering capability) of the plant over the year. In case the average actually achieved over the year is higher than the specified norm for plant availability, the generating company gets a higher payment. In case the average availability achieved is lower, the payment is also lower. Hence the name 'Availability Based Tariff'. This is the first component of Availability Based Tariff, and is termed 'capacity charge'.

The second component of Availability Based Tariff is the 'energy charge', which comprises of the variable cost (i.e., fuel cost) of the power plant for generating energy as per the given schedule for the day. It may specifically be noted that energy charge (at the specified plant-specific rate) is not based on actual generation and plant output, but on scheduled generation. In case there are deviations from the schedule (e.g., if a power plant delivers 600 MW while it was scheduled to supply only 500 MW), the energy charge payment would still be for the scheduled generation (500 MW), and the excess generation (100 MW) would get paid for at a rate dependent on the system conditions prevailing at the time. If the grid has surplus power at the time and frequency is above 50.0 cycles, the rate would be lower. If the excess generation takes place at the time of generation shortage in the system (in which condition the frequency would be below 50.0 cycles), the payment for extra generation would be at a higher rate.

To recapitulate, the Indian version of Availability Based Tariff comprises of three components: (a) capacity charge, towards reimbursement of the fixed cost of the plant, linked to the plant's declared capacity to supply MWs, (b) energy charge, to reimburse the fuel cost for scheduled generation, and (c) a payment for deviations from schedule, at a rate dependent on system conditions. The last component would be negative (indicating a payment by the generator for the deviation) in case the power plant is delivering less power than scheduled.

How do the Beneficiaries Share the Payments?

The Central generating stations in different regions of the country have various States of the Region as their specified beneficiaries or bulk consumers. The latter have shares in these plants calculated according to Gadgil formula, and duly notified by the Ministry of Power. The beneficiaries have to pay the capacity charge for these plants in proportion to their share in the respective plants. This payment is dependent on the declared output capability of the plant for the day and the beneficiary's percentage share in that plant, and not on power/energy intended to be drawn or actually drawn by the beneficiary from the Central station.

The energy charge to be paid by a beneficiary to a Central station for a particular day would be the fuel cost for the energy scheduled to be supplied from the power plant to the beneficiary during the day. In addition, if a beneficiary draws more power from the regional grid than what is totally scheduled to be supplied to him from the various Central generating stations at a particular time, he has to pay for the excess drawal at a rate dependent on the system conditions, the rate being lower if the frequency is high, and being higher if the frequency is low.

How does the Mechanism Work?

The process starts with the Central generating stations in the region declaring their expected output capability for the next day to the Regional Load Despatch Center (RLDC). The RLDC breaks up and tabulates these output capability declarations as per the beneficiaries' plant-wise shares and conveys their entitlements to State Load Despatch Centers (SLDCs). The latter then carry out an exercise to see how best they can meet the load of their consumers over the day, from their own generating stations, along with their entitlement in the Central stations. They also take into account the irrigation release requirements and load curtailment etc., that they propose in their respective areas. The SLDCs then convey to the RLDC their schedule of power drawal from the Central stations (limited to their entitlement for the day). The RLDC aggregates these requisitions and determines the despatch schedules for the Central generating stations and the drawal schedules for the beneficiaries duly incorporating any bilateral agreements and adjusting for transmission losses. These schedules are then issued by the RLDC to all concerned and become the operational as well as commercial datum. However, in case of contingencies, Central stations can prospectively revise the output capability declaration, beneficiaries can prospectively revise requisitions, and the schedules are correspondingly revised by RLDC.

While the schedules so finalised become the operational datum, and the regional constituents are expected to regulate their generation and consumer load in a way that the actual generation and drawls generally follow these schedules, deviations are allowed as long as they do not endanger the system security. The schedules are also used for determination of the amounts payable as energy charges, as described earlier. Deviations from schedules are determined in 15-minute time-blocks through special metering, and these deviations are priced depending on frequency. As long as the actual generation/drawal is equal to the given schedule, payment on account of the third component of Availability Based Tariff is zero. In case of under-drawal, a beneficiary is paid back to that extent according to the frequency dependent rate specified for deviations from schedule.

Why was Availability Based Tariff Necessary?

Prior to the introduction of Availability Based Tariff, the regional grids had been operating in a very undisciplined and haphazard manner. There were large deviations in frequency from the rated frequency of 50.0 cycles per second (Hz). Low frequency situations result when the total generation available in the grid is less than the total consumer load. These can be curtailed by enhancing generation and/or curtailing consumer load. High frequency is a result of insufficient backing down of generation when the total consumer load has fallen during off-peak hours. The earlier tariff mechanisms did not provide any incentive for either backing down generation during off-peak hours or for reducing consumer load/enhancing generation during peak-load hours. In fact, it was profitable to go on generating at a high level even when the consumer demand had come down. In other words, the earlier tariff mechanisms encouraged grid indiscipline.

The Availability Based Tariff directly addresses these issues. Firstly, by giving incentives for enhancing output capability of power plants, it enables more consumer load to be met during peak-load hours. Secondly, backing down during off-peak hours no longer results in financial loss to generating stations, and the earlier incentive for not backing down is neutralised. Thirdly, the shares of beneficiaries in the Central generating stations acquire a meaning, which was previously missing. The beneficiaries now have well-defined entitlements, and are able to draw power up to the specified limits at normal rates of the respective power plants. In case of over-drawal, they have to pay at a higher rate during peak-load hours, which discourages them from overdrawing further. This payment then goes to beneficiaries who received less energy than was scheduled, and acts as an incentive/compensation for them.

How does it Benefit Everyone?

The mechanism has dramatically streamlined the operation of regional grids in India. Firstly, through the system and procedure in place, constituents' schedules get determined as per their shares in Central stations, and they clearly know the implications of deviating from these schedules. Any constituent which helps others by under-drawal from the regional grid in a deficit situation, gets compensated at a good price for the quantum of energy under-drawn. Secondly, the grid parameters, i.e., frequency and voltage, have improved, and equipment damage correspondingly reduced. During peak load hours, the frequency can be improved only by reducing drawls, and necessary incentives are provided in the mechanism for the same. High frequency situation on the other hand, is being checked by encouraging reduction in generation during off-peak hours. Thirdly, because of clear separation between fixed and variable charges, generation according to merit-order is encouraged and pithead stations do not have to back down normally. The overall generation cost accordingly comes down. Fourthly, a mechanism is established for harnessing captive and cogeneration and for bilateral trading between the constituents. Lastly, Availability Based Tariff, by rewarding plant availability, enables more consumer load to be catered at any point of time.

Annexure A-3.3

Extracts from the 'Central Electricity Regulatory Commission (Terms & Conditions of Tariff) Regulations, 2009' (Issued by CERC on 19 January 2009)

13. Components of Tariff

(1) The tariff for supply of electricity from a thermal generating station shall comprise two parts, namely, capacity charge (for recovery of annual fixed cost consisting of the components specified to in regulation 14) and energy charge (for recovery of primary fuel cost and limestone cost where applicable).

(2) The tariff for supply of electricity from a hydro generating station shall comprise capacity charge and energy charge to be derived in the manner specified in regulation 22, for recovery of annual fixed cost (consisting of the components referred to in regulation 14) through the two charges.

(3) The tariff for transmission of electricity on inter-State transmission system shall comprise transmission charge for recovery of annual fixed cost consisting of the components specified in regulation 14.

14. Annual Fixed Cost

The annual fixed cost (AFC) of a generating station or a transmission system shall consist of the following components–

(a) Return on equity;

(b) Interest on loan capital;

(c) Depreciation;

(d) Interest on working capital;

(e) Operation and maintenance expenses;

(f) Cost of secondary fuel oil (for coal-based and lignite fired generating stations only);

(g) Special allowance in lieu of R&M or separate compensation allowance, wherever applicable.

21. Computation and Payment of Capacity Charge and Energy Charge for Thermal Generating Stations

(1) The fixed cost of a thermal generating station shall be computed on annual basis, based on norms specified under these regulations, and recovered on monthly basis under capacity charge. The total capacity charge payable for a generating station

shall be shared by its beneficiaries as per their respective percentage share/allocation in the capacity of the generating station.

(2) The capacity charge (inclusive of incentive) payable to a thermal generating station for a calendar month shall be calculated in accordance with the following formulae:

(a) Generating stations in commercial operation for less than ten (10) years on 1st April of the financial year:

AFC x (NDM/NDY) x (0.5 + 0.5 x PAFM/NAPAF) (in Rupees);

Provided that in case the plant availability factor achieved during a financial year (PAFY) is less than 70 per cent, the total capacity charge for the year shall be restricted to

AFC x (0.5 + 35/NAPAF) x (PAFY/70) (in Rupees).

(b) For generating stations in commercial operation for ten (10) years or more on 1st April of the year:

AFC x (NDM/NDY) x (PAFM/NAPAF) (in Rupees).

Where,

AFC = Annual fixed cost specified for the year, in Rupees.

NAPAF = Normative annual plant availability factor in percentage.

NDM = Number of days in the month.

NDY = Number of days in the year.

PAFM = Plant availability factor achieved during the month, in per cent.

PAFY = Plant availability factor achieved during the year, in per cent.

(3) The PAFM and PAFY shall be computed in accordance with the following formula:

$$\text{PAFM or PAFY} = 10000 \times \sum_{i=1}^{N} \text{DCi} / \{ \text{N} \times \text{IC} \times (100 - \text{AUX}) \} \%$$

Where,

AUX = Normative auxiliary energy consumption in percentage.

DCi = Average declared capacity (in ex-bus MW), subject to clause (4) below, for the ith day of the period i.e., the month or the year as the case may be, as certified by the concerned load despatch centre after the day is over.

IC = Installed Capacity (in MW) of the generating station.

N = Number of days during the period i.e., the month or the year as the case may be.

Note : DCi and IC shall exclude the capacity of generating units not declared under commercial operation. In case of a change in IC during the concerned period, its average value shall be taken.

(4) In case of fuel shortage in a thermal generating station, the generating company may propose to deliver a higher MW during peak-load hours by saving fuel during off-peak hours. The concerned Load Despatch Centre may then specify a pragmatic day-ahead schedule for the generating station to optimally utilise its MW and energy capability, in consultation with the beneficiaries. DCi in such an event shall be taken to be equal to the maximum peak-hour ex-power plant MW schedule specified by the concerned Load Despatch Centre for that day.

(5) The energy charge shall cover the primary fuel cost and limestone consumption cost (where applicable), and shall be payable by every beneficiary for the total energy scheduled to be supplied to such beneficiary during the calendar month on ex-power plant basis, at the energy charge rate of the month (with fuel and limestone price adjustment). Total Energy charge payable to the generating company for a month shall be:

(Energy charge rate in Rs/kWh) x {Scheduled energy (ex-bus) for the month in kWh.}

(6) Energy charge rate (ECR) in Rupees per kWh on ex-power plant basis shall be determined to three decimal places in accordance with the following formulae :

(a) For coal based and lignite fired stations

ECR = { (GHR – SFC x CVSF) x LPPF/CVPF + LC x LPL } x 100/(100 – AUX)

(b) For gas and liquid fuel based stations

ECR = GHR x LPPF x 100/{CVPF x (100 – AUX) }

Where,

AUX = Normative auxiliary energy consumption in percentage.

CVPF = Gross calorific value of primary fuel as fired, in kCal per kg, per litre or per standard cubic metre, as applicable.

CVSF = Calorific value of secondary fuel, in kCal per ml.

ECR = Energy charge rate, in Rupees per kWh sent out.

GHR = Gross station heat rate, in kCal per kWh.

LC = Normative limestone consumption in kg per kWh.

LPL = Weighted average landed price of limestone in Rupees per kg.

LPPF = Weighted average landed price of primary fuel, in Rupees per kg, per litre or per standard cubic metre, as applicable, during the month.

SFC = Specific fuel oil consumption, in ml per kWh.

(7) The landed cost of fuel for the month shall include price of fuel corresponding to the grade and quality of fuel inclusive of royalty, taxes and duties as applicable, transportation cost by rail/road or any other means, and, for the purpose of computation of energy charge, and in case of coal/lignite shall be arrived at after considering normative transit and handling losses as percentage of the quantity of coal or lignite dispatched by the coal or lignite supply company during the month as given below :

Pithead generating stations: 0.2 per cent

Non-pithead generating stations: 0.8 per cent

(8) The landed price of limestone shall be taken based on procurement price of limestone for the generating station, inclusive of royalty, taxes and duties as applicable and transportation cost for the month.

(9) The tariff structure as provided in this regulation may be adopted by the Department of Atomic Energy, Government of India for the nuclear generating stations by specifying annual fixed cost (AFC), normative annual plant availability factor (NAPAF), installed capacity (IC), normative auxiliary power consumption (AUX) and energy charge rate (ECR) for such stations.

22. Computation and Payment of Capacity Charge and Energy Charge for Hydro Generating Stations

(1) The fixed cost of a hydro generating station shall be computed on annual basis, based on norms specified under these regulations, and recovered on monthly basis under capacity charge (inclusive of incentive) and energy charge, which shall be payable by the beneficiaries in proportion to their respective allocation in the saleable capacity of the generating station, that is to say, in the capacity excluding the free power to the home State:

> Provided that during the period between the date of commercial operation of the first unit of the generating station and the date of commercial operation of the generating station, the annual fixed cost shall provisionally be worked out based on the latest estimate of the completion cost for the generating station, for the purpose of determining the capacity charge and energy charge payment during such period.

(2) The capacity charge (inclusive of incentive) payable to a hydro generating station for a calendar month shall be:

AFC x 0.5 x NDM/NDY x (PAFM/NAPAF) (in Rupees)

Where,

AFC = Annual fixed cost specified for the year, in Rupees.

NAPAF = Normative plant availability factor in percentage.

NDM = Number of days in the month.

NDY = Number of days in the year.

PAFM = Plant availability factor achieved during the month, in percentage.

(3) The PAFM shall be computed in accordance with the following formula :

$$PAFM = 10000 \times \sum_{i=1}^{N} DCi / \{ N \times IC \times (100 - AUX) \} \%$$

Where,

AUX = Normative auxiliary energy consumption in percentage.

DCi = Declared capacity (in ex-bus MW) for the ith day of the month which the station can deliver for at least three (3) hours, as certified by the nodal load despatch centre after the day is over.

IC = Installed capacity (in MW) of the complete generating station.

N = Number of days in the month.

(4) The energy charge shall be payable by every beneficiary for the total energy scheduled to be supplied to the beneficiary, excluding free energy, if any, during the calendar month, on ex-power plant basis, at the computed energy charge rate. Total Energy charge payable to the generating company for a month shall be:

(Energy charge rate in Rs/kWh) x {Scheduled energy (ex-bus) for the month in kWh} x

(100 – FEHS)/100.

(5) Energy charge rate (ECR) in Rupees per kWh on ex-power plant basis, for a hydro generating station, shall be determined up to three decimal places based on the following formula, subject to the provisions of clause (7):

$$ECR = AFC \times 0.5 \times 10/\{ DE \times (100 - AUX) \times (100 - FEHS)\}$$

Where,

DE = Annual design energy specified for the hydro generating station, in MWh, subject to the provision in clause (6) below.

FEHS = Free energy for home State, in per cent, as defined in regulation 32.

(6) In case actual total energy generated by a hydro generating station during a year is less than the design energy for reasons beyond the control of the generating company, the following treatment shall be applied on a rolling basis:

(i) in case the energy shortfall occurs within ten years from the date of commercial operation of a generating station, the ECR for the year following the year of energy shortfall shall be computed based on the formula specified in clause (5) with the modification that the DE for the year shall be considered as equal to the actual energy generated during the year of the shortfall, till the energy charge shortfall of the previous year has been made up, after which normal ECR shall be applicable;

(ii) in case the energy shortfall occurs after ten years from the date of commercial operation of a generating station, the following shall apply:

Suppose the specified annual design energy for the station is DE MWh, and the actual energy generated during the concerned (first) and the following (second) financial years is A1 and A2 MWh respectively, A1 being less than DE. Then, the design energy to be considered in the formula in clause (5) of this Regulation for calculating the ECR for the third financial year shall be moderated as (A1 + A2 – DE) MWh, subject to a maximum of DE MWh and a minimum of A1 MWh.

(iii) Actual energy generated (e.g., A1, A2) shall be arrived at by multiplying the net metered energy sent out from the station by 100/(100 – AUX).

(7) In case the energy charge rate (ECR) for a hydro generating station, as computed in clause (5) above, exceeds eighty paise per kWh, and the actual saleable energy in a year exceeds

{ DE x (100 – AUX) x (100 – FEHS)/10000 } MWh, the Energy charge for the energy in excess of the above shall be billed at eighty paise per kWh only:

Provided that in a year following a year in which total energy generated was less than the design energy for reasons beyond the control of the generating company, the energy charge rate shall be reduced to eighty paise per kWh after the energy charge shortfall of the previous year has been made up.

(8) The concerned Load Despatch Center shall finalise the schedules for the hydro generating stations, in consultation with the beneficiaries, for optimal utilisation of all the energy declared to be available, which shall be scheduled for all beneficiaries in proportion to their respective allocations in the generating station.

24. Unscheduled Interchange (UI) Charges

(1) All variations between actual net injection and scheduled net injection for the generating stations, and all variations between actual net drawal and scheduled net drawal for the beneficiaries shall be treated as their respective Unscheduled Interchanges (UI), charges for which shall be governed by the relevant regulations specified by the Commission from time to time.

(2) Actual net unscheduled interchange of every inter-State entity shall be metered on its periphery through special energy meters (SEMs) installed by the Central Transmission Utility (CTU), and computed in MWh for each 15-minute time-block by the concerned Regional Load Despatch Center.

11. Sale of Infirm Power

Supply of infirm power shall be accounted as Unscheduled Interchange (UI) and paid for from the regional or State UI pool account at the applicable frequency-linked UI rate.

Provided that any revenue earned by the generating company from sale of infirm power after accounting for the fuel expenses shall be applied for reduction in capital cost.

3. Definitions

In these regulations, unless the context otherwise requires:

(20) 'infirm power' means electricity injected into the grid prior to the commercial operation of a unit or block of the generating station.

4

Analysis of Competition and Market Power in Wholesale-Electricity Market in India*

Umesh Kumar Shukla and
Ashok Thampy

Introduction

The electricity market reform is believed to bring higher efficiency, lower and more stable prices, reliable services and timely investment, but the outcomes of the deregulation in electricity markets are mixed (Besant-Jones and Tenenbaum, 2001; Brennan *et al.*, 2002; Woo *et al.*, 2003). As a part of electricity reform in India, the Electricity Act, 2003 aims to promote competition while protecting the interest of the consumers. National Electricity Policy and Tariff Policy notified by the Central government in February 2005 and January 2006 respectively also aims to ensure availability of electricity at reasonable and competitive prices. The Central Electricity Regulatory Commission (CERC) has also taken various initiatives to facilitate the competition in the electricity market, which includes Availability Based Tariff (ABT) in 2003, open access in inter-state transmission in 2004, regulations for grant of trading licence in 2004, Indian Electricity Grid Code, guidelines for grant of permission for setting up and operation of power exchanges in 2007 and grant of approval to set up the power exchanges in 2008 to Indian Energy Exchange Limited (IEXL) and Power Exchange India Limited (PXIL). One of the main objectives for starting of power exchanges was to promote the competition and bring

---

* The research paper was presented at 5th *International Conference on Public Policy and Management* organised by Indian Institute of Management, Bangalore, during 8-11 August 2010.

The views expressed in this paper are those of the authors and do not necessarily reflect the views of the organisations/departments of the authors.

down the electricity prices in the wholesale electricity market in India (WEMI). This paper examines short-term transactions during July 2008 to September 2009 to understand the impact of power exchanges on the electricity prices in WEMI. Section II examines actual demand-supply position in different regions of India, inter-regional transfer capacity of electricity, and legal framework of trading and discusses about different trading mechanisms of WEMI. Section III analyses the impact of first power exchange, IEXL, on the electricity prices of bilateral contracts and UI rate as well as on the grid frequency. It also analyses the impact of second power exchange, PXIL, on the electricity prices of bilateral contracts and UI rate and the impact on the prices in the power exchanges. Last Section suggests various measures in the context of the findings of the study.

Electricity Market

Demand-Supply Position

The growth in installed capacity and generation of electricity in India has not been sufficient for increasing demand resulting into substantial peak and energy shortages in the country. During the year 2008-09, the peak and energy shortages of electricity were 11.95 per cent and 11.02 per cent respectively. Moreover, actual power supply position in different regions of India also varied as shown in Table 4.1.

As the electricity requirement fluctuates during different time of the day, there might be surpluses in some locations at some times despite an overall shortage of electricity. For example, in hilly regions, the requirement of power increases during winter season and decreases in summer, while in southern and western regions, it decreases in winter and increases in summer. Similarly, the period of peak during the day is also different in different regions. For example, there is normally difference of 2-3 hours in peak demand load between eastern/ north eastern regions and rest of the country. Thus trading opportunities in India exist across time in view of seasonal, time-of-day (peak demand time varies in cities/states) and unanticipated demand variations.

Table 4.1

Actual Power Supply Position in 2008-09 in Different Regions of India

Region	*Peak Demand (GW)*	*Peak Supply (GW)*	*Peak Shortages*		*Energy Demand (GWh)*	*Energy Supply (GWh)*	*Energy Shortages*		*Occurrence of Peak*
			(GW)	*(%)*			*(GWh)*	*(%)*	
Northern	33.034	29.504	3.530	10.69	224,218	199,928	24,290	10.83	June, July and August
Western	37.240	30.154	7.086	19.03	254,486	213,724	40,762	16.02	December
Southern	28.340	26.244	2.096	7.40	204,086	188,865	15,221	7.46	March and October
Eastern	12.901	11.689	1.212	9.39	82,127	78,370	3,757	4.57	March
North Eastern	1.820	1.358	0.462	25.38	9,407	8,134	1,273	13.53	January
All India	109.809	96.685	13.124	11.95	774,324	689,021	85,303	11.02	March and October

Source: Central Electricity Authority (2009).

Transmission System

In order to facilitate the transfer of elcctricity between neighbouring states, state grids are inter-connected through high-voltage transmission links to form a regional grid. There are five regional grids:

1. *Northern region:* Delhi, Haryana, Himachal Pradesh, Jammu and Kashmir, Punjab, Rajasthan, Uttarakhand and Uttar Pradesh.
2. *Eastern region:* Bihar, Jharkhand, Orissa, Sikkim and West Bengal.
3. *Western region:* Dadra and Nagar Haveli, Daman and Diu, Chhattisgarh, Goa, Gujarat, Madhya Pradesh and Maharashtra.
4. *Southern region:* Andhra Pradesh, Karnataka, Kerala, Puducherry and Tamil Nadu.
5. *North Eastern region:* Arunachal Pradesh, Assam, Manipur, Meghalaya, Mizoram, Nagaland and Tripura.

By the end of March 2009, there was inter-regional transfer capacity of 20,750 MW between the regions. Ministry of Power has envisaged establishment of an integrated national power grid in the country by 2012 with an inter-regional transfer capacity of 38,650 MW (Ministry of Power, 2009). The Indian T&D system is characterised by exceptionally high losses of about 27 per cent, as compared to the losses of 10-15 per cent in developed countries.

Legal Framework

The Electricity Act, 2003 recognised the transaction involving purchase and sale of electricity as a distinct licensed activity termed as 'trading'. The electricity regulatory commissions have powers to grant the trading licence—the CERC for inter-state trading and the SERCs for intra-state trading. While there is no provision for any tariff fixation for trading, electricity regulatory commissions may fix trading margin, if considered necessary. CERC has notified the regulations in 2004 prescribing procedure, terms and conditions for grant of inter-state trading licence to provide enabling framework for bilateral trading, between generator and trader and trader and licensee. The Electricity Act, 2003 has vested the responsibility of developing the market in electricity with the government and the electricity regulatory commission. In line with the responsibility under the Act, CERC issued the guidelines in February 2007 for grant of permission for setting up and operation of power exchange.

Trading Mechanism

There are four trading mechanisms used in WEMI.

Long-Term Power Purchase Agreements

While, the suppliers are mostly the Central or state-owned generating stations and a few individual power producers (IPPs), the buyers are generally the state electricity boards (SEBs)/distribution companies (Discoms). The power from various generating stations is allocated to SEBs/ Discoms and the appropriate commission regulates the price on the basis of normative historical cost based on the tariff principles notified by the regulatory commission or as per the PPA. The SEBs/Discoms, which have the obligation to provide electricity to their consumers, mainly rely on supplies from these long-term contracts. The annual transmission charges determined, after deducting adjustable revenue from the short-term customers, are shared by the long-term customers.

Short-Term Bilateral Contracts

The concept of short-term trading through bilateral contracts was introduced in 2001 by PTC India Limited as a second mechanism to trade short-term surplus of power. Short-term trading contracts are mostly inter-state or inter-regional, requiring open access through the central transmission utility (CTU) network and purchase and sale of electricity takes place mostly through the traders, which have been granted licence by the CERC. A couple of years ago, in the initial phase of electricity trading, the prices were settled through mutual negotiations. Nowadays the sellers invite bids and the trader with highest bid price is selected, which in turn, sells the electricity to a buyer after adding trading margin. Figure 4.1 shows that the trading volume through short-term bilateral contracts recorded more than 13-fold increases from 1,617 GWh in 2001-02 to 21,917 GWh in 2008-09, and as a result the share of traded volume through short-term bilateral contracts to the total electricity generation in India has increased from 0.31 per cent in 2001-02 to 3.03 per cent in 2008-09.

Figure 4.2 shows the weighted average sale price of electricity traded through bilateral contracts during the years 2004-05 to 2008-09:

Figure 4.1

Traded Volume through Short-Term Bilateral Contracts and Total Electricity Generation in India

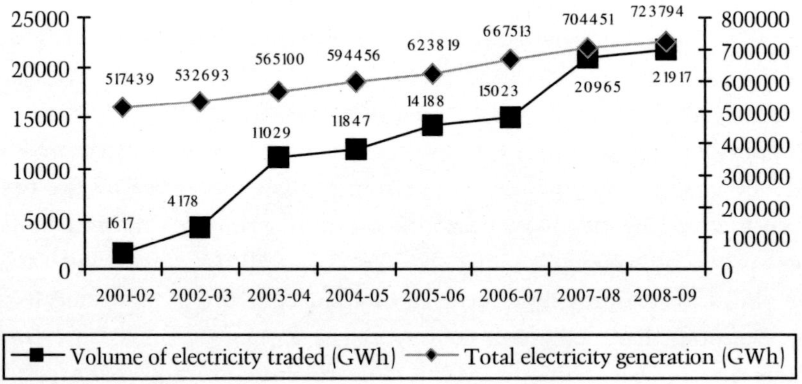

Note: Based on the data of trading through bilateral contracts compiled from the website of CERC, *www.cercind.gov.in* accessed in August 2009 and *Economic Survey, 2008-09*.

Source: CERC and Ministry of Finance.

Figure 4.2

Price Range and Volume of Electricity Traded Over the Years

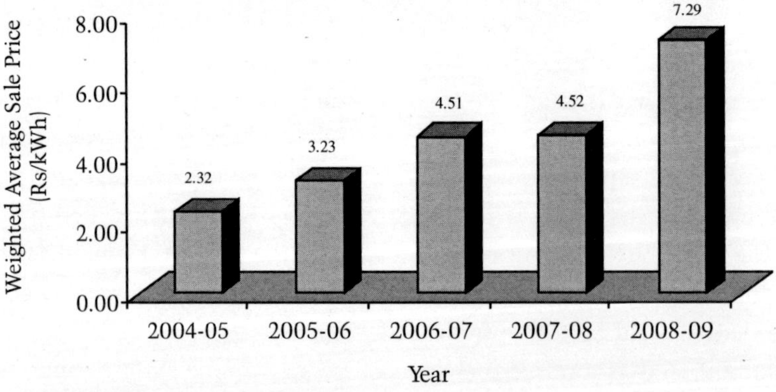

Note: Based on the data of trading through bilateral contracts compiled from the website of CERC, *www.cercind.gov.in* accessed in August 2009 and *Economic Survey, 2008-09*.

Source: CERC and Ministry of Finance.

Unscheduled Interchange

With the introduction of ABT in 2003, unscheduled interchange (UI) became the third mechanism setting the benchmark price for trading of electricity. The ABT[1] comprises three components: (a) capacity charge, towards reimbursement of the fixed cost of the plant, linked to the plant's declared capacity of electricity supply, (b) energy charge, to reimburse the fuel cost for scheduled generation and (c) a payment for deviations from the schedule, at a rate dependent on system conditions. The variation between actual generation or actual drawl and scheduled generation or scheduled drawl is accounted for through UI charges. UI for a generating station or a seller is equal to its actual generation minus its scheduled generation. UI for a beneficiary or buyer is equal to its total actual drawl minus its total scheduled drawl. For calculation of UI, the energy metered is compared with the scheduled energy in each 15-minute time-block, and the difference (+ or -) becomes the UI. The corresponding UI rate is determined by taking the average frequency for the same 15-minute time-block. All UI payments are made into and from a regional UI pool account, operated by the concerned RLDC.

Figure 4.3

UI Rate Over the Years

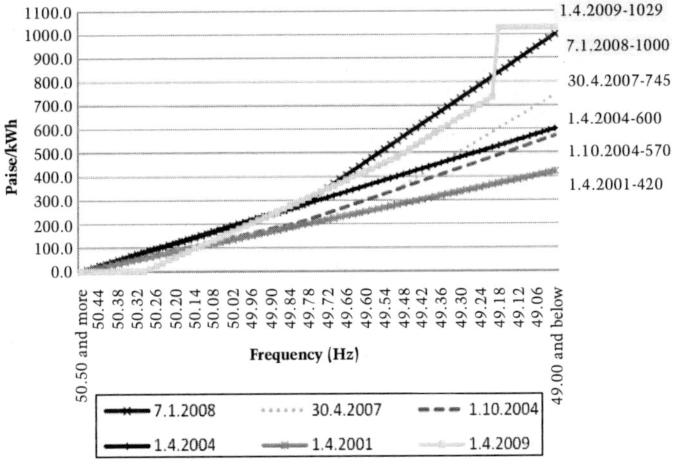

Note: Based on the data of UI rates compiled from various tariff and UI regulations available at the website of CERC, *www.cercind.gov.in* accessed in December 2009.

Source: CERC.

1. For details see Bhushan (2005).

Power Exchange

To develop the power market, CERC, in February 2007, issued the guidelines[2] for grant of permission for setting up and operation of power exchanges and accorded approval in August 2007 to set up a power exchange named Indian Energy Exchange Limited (IEXL). With the start of trading in IEXL on 27 June 2008, power exchange became the fourth mechanism. Subsequently CERC in September 2008 accorded approval for setting up another power exchange named Power Exchange India Limited (PXIL), which commenced its operations on 22 October 2008.[3] In the power exchange, trading takes place through collective transaction.[4] On a daily basis, the IEXL and PXIL offer a double-side closed auction for delivery on the following day (termed as day-ahead market). Price discovery is made through double-side bidding and buyers and suppliers pay/receive the uniform price. Market-splitting methodology[5] is adopted for congestion management. Each electrical region of the country has been divided in two bid areas to accommodate any exigencies of congestion in intra-regional transmission system. In case of congestion, the area prices would differ in each region after splitting the market and the members in different bid areas would pay different area prices. The funds collected due to differential area prices are kept in a separate fund to be utilised as per direction of

2. As per the guidelines, a company registered under the Companies Act, 1956, or a consortium of companies is eligible to apply for setting up a power exchange, if it has adequate knowledge and understanding of Indian Electricity Grid Code (IEGC), open access issues, ABT, UI mechanism, scheduling despatch and energy accounting procedure. The power exchange would have the features of de-mutualised form of organisation; reliable, effective and impartial management; ring-fencing between ownership, management and participation; transparency in operation and decision-making; computerised trading and clearing system; efficient clearing settlement and guarantee system and effective trade information dissemination system.
3. Subsequently, CERC accorded permission to set up third power exchange named National Power Exchange Limited (NPEX).
4. Central Electricity Regulatory Commission (Open Access for Inter-State Transmission) Regulations, 2008 defines the collective transaction as a set of transactions discovered in power exchange through anonymous, simultaneous competitive bidding by buyers and sellers.
5. If the contractual flow does not exceed the transmission capacity, a common price is established for the whole area. If the flow exceeds the capacity at the common price for the whole market area, it is split in surplus and deficit part and the price is reduced in the surplus area (sale>purchase) and increased in the deficit area (purchase>sale). This would reduce the sale and increase the purchase in the surplus area and reduce the purchase and increase the sale in the deficit area and needed flow is reduced to match the available transfer capability.

CERC. All buyers and sellers bear transmission losses[6] and charges[7] in their regional transmission system till the periphery of the transmission system and are also required to bear the operating charges[8] to cover the costs of system operators and transaction fee of the exchange. After the power is scheduled, if the actual generation or actual drawl deviates from the schedule, the deviation is settled through UI mechanism. During the period August 2008 to September 2009, the maximum electricity price was more than Rs 8.00/kWh in all the months and has touched Rs 17.00/kWh in IEXL and Rs 15.00/kWh in PXIL during August 2009.

Impact of Power Exchanges on Electricity Prices

It is generally considered that power exchanges facilitate competition by providing transparency in the transactions between the buyers and sellers of electricity. This necessitates the need to analyse the impact of power exchanges on the electricity prices in WEMI. As price under long-term PPAs, determined on cost-plus basis, is not dependent on demand and supply, the possibility of market power exists in remaining three mechanisms. Therefore analysis has been made for the transactions of electricity transacted in WEMI through short-term bilateral contracts, UI and power exchanges. First the impact of IEXL on electricity prices in WEMI have been analysed on other two mechanisms of short-term transactions (*viz.*, trading through short-term bilateral contracts and UI) and on the grid frequency. Thereafter the impact of PXIL has been analysed on electricity prices in WEMI on these two mechanisms as well as prices in other exchange.

---

6. Transmission losses are notional losses of electricity, when it passes through a transmission network and is required to be borne by the buyer or seller. The seller would absorb the transmission losses from the point of injection till the delivery point and the buyer would absorb the transmission losses from the delivery point till the point of drawl. In case of collective transactions, the delivery point is respective regional periphery. However, in a bilateral transaction, it is mutually agreed. Thus, the buyer would draw less than contracted power and the seller would need to inject more than contracted power.
7. Transmission charges are payable to the transmission and distribution licensee(s) for the use of their transmission and distribution system(s) to be paid by the seller and buyer.
8. Operating charges are the charges payable to the system operators i.e. SLDCs, RLDCs and NLDC for scheduling and reserving the transmission capacity for short-term open access transactions.

Impact of IEXL

Figure 4.4 shows daily volume and weighted average prices in IEXL during the period July-September 2008.

Figure 4.4

Daily Volume and Electricity Prices in IEXL during July-September 2008

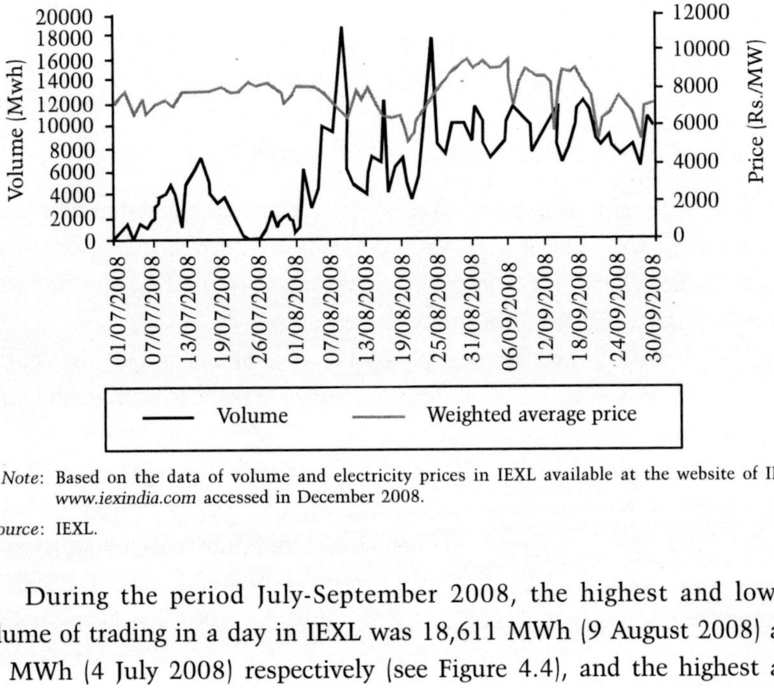

Note: Based on the data of volume and electricity prices in IEXL available at the website of IEXL, *www.iexindia.com* accessed in December 2008.

Source: IEXL.

During the period July-September 2008, the highest and lowest volume of trading in a day in IEXL was 18,611 MWh (9 August 2008) and 74 MWh (4 July 2008) respectively (see Figure 4.4), and the highest and lowest price was Rs 11,001/MWh (between 19-20 hours on 18 September 2008) and Rs 922/MWh (between 04-05 hours on 10 August 2008). Figure 4.5 shows the spread in the daily high and low prices of electricity in IEXL.

Impact on Electricity Prices through Bilateral Contracts

The weighted average price of electricity during July-September 2008 in IEXL were Rs 7.78/kWh as compared to Rs 6.91/kWh for electricity traded through bilateral contracts and was 13 per cent higher than bilateral contracts. It may be observed from Table 4.2 that the volume of electricity traded through bilateral contracts in July-September 2008 has gone down by 13 per cent as compared to the same quarter of the last year but has

Figure 4.5

Daily Spread between High and Low Electricity Prices in IEXL during July-September 2008

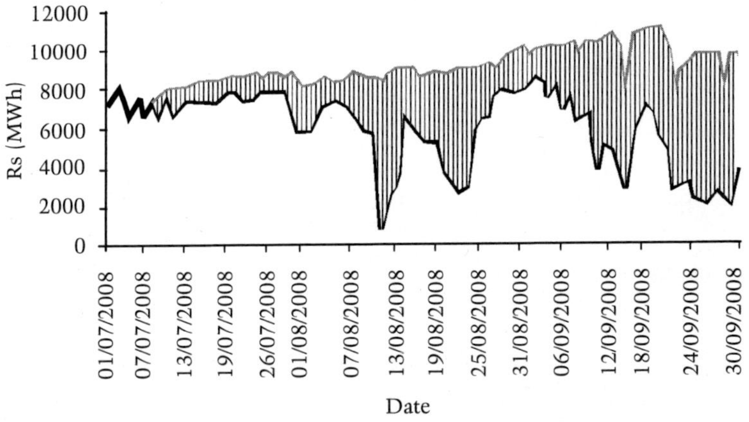

Note: * Based on the data of electricity prices in IEXL available at the website of IEXL, *www.iexindia.com* accessed in December 2008.

Source: IEXL.

increased by 41 per cent as compared with the immediately preceding quarter. The prices of electricity in bilateral contracts have shown substantial increase and have gone up by 105 per cent in July-September 2008 as compared to the same quarter of the last year but when compared to the prices of the immediately preceding quarter of April-June 2008 have gone down by 4 per cent.

Table 4.2

Comparison of Electricity Prices of Trading through Bilateral Contracts

Details	*July-September 2007*	*April-June 2008*	*July-September 2008*	*% Change from July-September 2007*	*% Change from April-June 2008*
Volume	7226.51	4451.78	6274.22	-13	41
Weighted average price	3.37	7.20	6.91	105	-4

Note: Based on the data of trading through bilateral contracts compiled from the website of CERC, *www.cercind.gov.in* accessed in August 2009.

Source: CERC.

However, comparison of electricity prices in bilateral contracts for the quarter April-June 2007 with July-September, 2007 shows that the reduction in prices in July-September quarter was 27 per cent in 2007 as compared to 4 per cent in 2008, which indicate that the reduction in prices in 2008 in July-September quarter has been lower by 23 per cent or in other words the prices have actually been higher. This indicates the possibility of use of market power by the sellers in view of shortage situation of electricity in the country. However, it does not lead us to a definitive conclusion as the trading through bilateral contracts and power exchanges are used to meet different nature of demand in the market.

Impact on UI Rate

The demand met through power exchanges is more comparable with the UI. Therefore, the impact of IEXL on the UI rate and volume has been analysed by using the data for the quarter April-June 2008, July-September 2008 and July-September 2007 of the western region. It may be seen from Figure 4.6 that weighted average daily UI rate in July-September 2008 (after starting of IEXL) have generally been higher than that in July-September 2007 (similar period of the last year).

Figure 4.6

Weighted Average Daily UI Rate of July-September Quarter of the Year 2008 and 2007 in Western Region

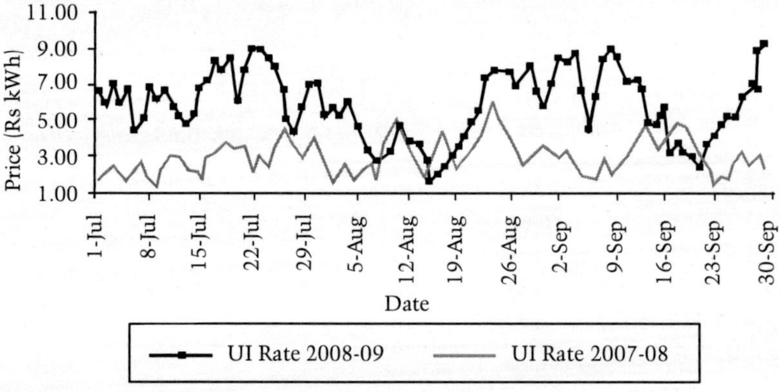

Note: Based on the data of UI rates compiled from the website of Western Regional Power Committee accessed in August 2009.

Source: Western Regional Power Committee.

It may be observed from Table 4.3 that weighted average daily UI rate during July-September 2008 is 79 per cent higher than that in July-September 2007. The standard deviation is also higher during July-September 2008 than July-September 2007 showing more volatility in the UI rate. Although weighted average daily UI rate during July-September 2008 is only 16 per cent higher than that in April-June 2008 but even this increase is alarming as there is normally more demand of electricity during the period April-June.

Table 4.3

Weighted Average Daily UI Rate of July-September 2008 and July-September 2007 in Western Region

	July-September 2007	*April-June 2008*	*July-September 2008*	*% Change from July-September 2007*	*% Change from April-June 2008*
Highest	6.03	9.25	9.12	51	-1
Lowest	1.53	1.60	1.76	15	10
Average	3.06	5.30	5.82	90	10
Weighted average	2.99	4.63	5.35	79	16
Standard deviation	0.97	2.24	1.90	96	-15

Note: Based on the data of UI rates compiled from the website of western regional power committee accessed in August 2009.

Source: Western Regional Power Committee.

Relationship of UI Rate with Electricity Prices in IEXL

As UI and power exchanges meet the similar nature of demand, there is possibility of inter-relationship between them. Therefore, analysis have been made to identify the relationship between weighted average daily UI rate and weighted average daily electricity prices in IEXL using the data of the western region for the quarter July-September 2008. It may be seen from Figure 4.7 that UI rate has normally been lower than the electricity prices in IEXL.

It may be seen from Table 4.4 that the electricity prices in IEXL were higher than UI rate. While the average and weighted average electricity prices in IEXL have been higher by about 37 per cent and 51 per cent respectively than UI rate, lowest electricity price was higher by about 180 per cent. However, the higher standard deviation of UI rate shows more fluctuation of the grid frequency.

Figure 4.7

UI Rate and Electricity Prices in IEXL during July-September 2008 in Western Region

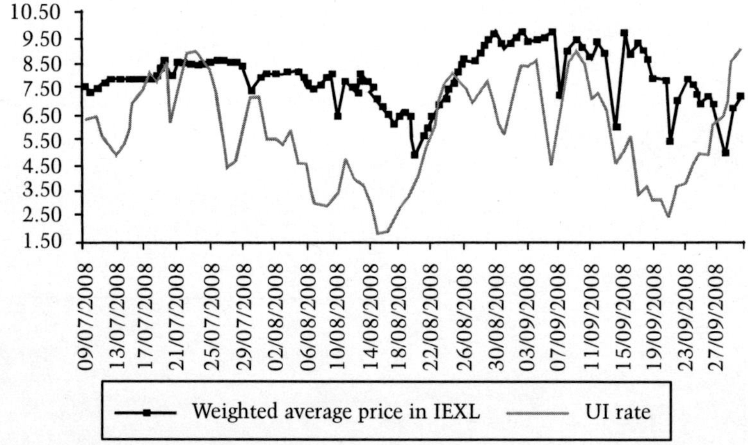

Note: Based on the data of UI rates compiled from the website of Western Regional Power Committee and electricity prices in IEXL available at the website of IEXL, *www.iexindia.com* accessed in August 2009.

Source: Western Regional Power Committee and IEXL.

Table 4.4

Comparison of Electricity Prices in IEXL and UI during July-September 2008 in Western Region

Particulars	*Electricity Prices in IEXL*	*UI Rate*	*Higher Price in Power Exchange (%)*
Highest price	9.80	9.12	7
Lowest price	4.92	1.76	180
Average price	7.95	5.82	37
Weighted average price	8.07	5.35	51
Standard deviation	1.09	1.98	

Note: Based on the data of UI rates compiled from the website of Western Regional Power Committee and electricity prices in IEXL available at the website of IEXL *www.iexindia.com* accessed in August 2009.

Source: Western Regional Power Committee and IEXL.

It may be observed from Table 4.5 that there is positive correlation of 0.4329 between the electricity prices in IEXL ($IEXL_t$) and UI rate (UI_t), suggesting that price formation in IEXL is positively linked to the UI rate.

The correlation decreases with $IEXL_{t-1}$ and UI_t, $IEXL_{t-2}$ and UI_t, and $IEXL_{t-3}$ and UI_t and so on but increases with $IEXL_{t+1}$ and UI_t, $IEXL_{t+2}$ and UI_t and $IEXL_{t+3}$ and UI_t and so on upto $IEXL_{t+4}$. This indicates that UI rate has the impact on the electricity prices in IEXL and that impact is more visible after 3-4 days. This also indicates the possibility of overdrawing the power from the grid by paying UI charges to sell that power in the IEXL at higher prices by market participants by using the UI mechanism to make the profit.

Table 4.5

Correlation of Electricity Prices in IEXL and UI Rate during July-September 2008 in Western Region

CORREL IEXLt, UI(t)	0.4329		
CORREL IEXL(t-1), UI(t)	0.2481	CORREL IEXL(t+1), UI(t)	0.5274
CORREL IEXL(t-2), UI(t)	0.1382	CORREL IEXL(t+2), UI(t)	0.6270
CORREL IEXL(t-3), UI(t)	0.1326	CORREL IEXL(t+3), UI(t)	0.6478
CORREL IEXL(t-4), UI(t)	0.0637	CORREL IEXL(t+4), UI(t)	0.6505
CORREL IEXL(t-5), UI(t)	0.0048	CORREL IEXL(t+5), UI(t)	0.6268

Note: Where IEXL=IEXL price, UI=UI rate, t=same day, $_{t+1}$=next day, $_{t+2}$=one day after next day and so on. Similarly $_{t-1}$=previous day, $_{t-2}$=one day previous to previous day and so on.

It may be observed from Table 4.6 that there is negative correlation of 0.4049 between volume of electricity traded in IEXL ($IEXLV_t$) and UI (UI_t). The correlation decreases with $IEXLV_{t-1}$ and UIV_t, $IEXLV_{t-2}$ and UIV_t, and $IEXLV_{t-3}$ and UIV_t and so on but increases with $IEXLV_{t+1}$ and UIV_t and $IEXLV_{t+2}$ and UIV_t. This indicates that UI has the impact on the volume of electricity traded in IEXL and that impact is more visible after two days.

Table 4.6

Correlation of UI and Net Buy Volume in IEXL during July-September 2008 in Western Region

CORREL IEXLV(t), UIV(t)	-0.4049		
CORREL IEXLV(t-1), UIV(t)	-0.3423	CORREL IEXLV(t+1), UIV(t)	-0.4715
CORREL IEXLV(t-2), UIV(t)	-0.2788	CORREL IEXLV(t+2), UIV(t)	-0.5083
CORREL IEXLV(t-3), UIV(t)	-0.2046	CORREL IEXLV(t+3), UIV(t)	-0.4585
CORREL IEXLV(t-4), UIV(t)	-0.0963	CORREL IEXLV(t+4), UIV(t)	-0.3631
CORREL IEXLV(t-5), UIV(t)	-0.0595	CORREL IEXLV(t+5), UIV(t)	-0.2794

Note: Where IEXLV=IEXL volume and UIV=UI and t=same day, $_{t+1}$=next day, $_{t+2}$=one day after next day and so on. Similarly, $_{t-1}$=previous day, $_{t-2}$=one day previous to previous day and so on.

Impact on Grid Frequency

The higher standard deviation of UI rate and significant correlation between volumes of electricity traded in IEXL and UI indicate fluctuation in the grid frequency. This has great concern for the grid security in view of exorbitant cost associated with the grid disturbances.[9] Based on the weighted average daily UI rate of the western region, grid frequency during the period July-September 2008 and July-September 2007 has been worked out, which indicates that grid frequency has generally been lower in July-September 2008 as compared to July-September 2007 as shown in Figure 4.8.

Figure 4.8

Average Grid Frequency during July-September in 2008 and 2007 in Western Region

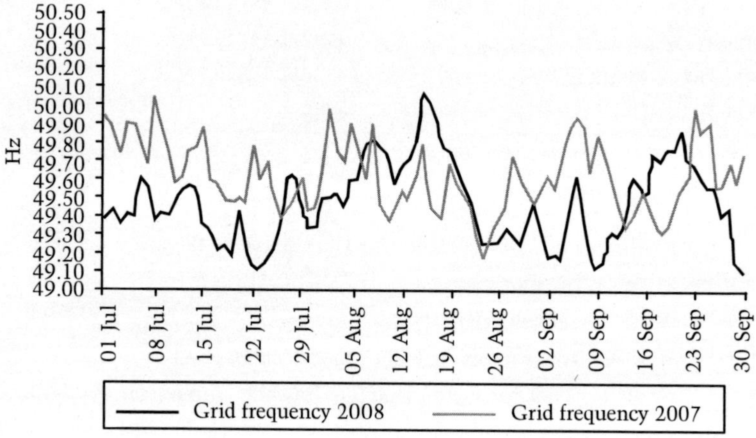

It may be seen from Table 4.7 that the average grid frequency during July-September 2008 has been lower at 49.47 Hz as compared to 49.62 Hz in July-September 2007. The standard deviation in grid frequency during July-September 2008 has also been higher at 0.22 as compared to 0.19 in July-September 2007 showing more fluctuations of the grid frequency. Thus, the impact of IEXL on the grid has seen greater fluctuations in the grid frequency.

9. It is estimated that a major grid disturbance costs about 130-180 billion US Dollar, while a minor grid disturbance costs 20 million US Dollar (PGCIL website accessed in May 2008).

Table 4.7

Average Grid Frequency during July-September in 2008 and 2007 in Western Region

Month	*Average Grid Frequency*		*Standard Deviation*	
	2007	*2008*	*2007*	*2008*
July	49.66	49.38	0.18	0.14
August	49.57	49.58	0.19	0.23
September	49.62	49.44	0.19	0.23
July-September	49.62	49.47	0.19	0.22

Impact of Start of PXIL

The data of electricity prices and volume of trading in IEXL and PXIL for the period commencing from the start of exchanges till quarter ending September 2009, available on the websites of IEXL and PXIL and data of monthly reports on short-term transactions of electricity for the period August 2008 to September 2009, available on CERC website, are used for analysing the impact. The gap between the volume of buy-bids and sale-bids placed through power exchange shows that there was more demand (1.04 times) when compared with the supply offered through IEXL and less demand (0.87 times) when compared with the supply offered through PXIL (Ministry of Power, 2009).

Impact on Electricity Prices of Bilateral Contracts and UI Rate

It may be seen from Figure 4.9 that the wholesale electricity prices in power exchanges (IEXL and PXIL) have generally been higher than UI rates and short-term bilateral transactions through traders.

It may also be observed that after start of PXIL, the electricity prices in short-term bilateral transactions through traders and UI rates have shown the downward trend. This may be due to the reason that power exchange, in view of the transparency in transactions, is perceived as the preferred option for purchase and sell of electricity, which is evident from the increasing share of power exchanges in short-term transactions of electricity as shown in Table 4.8.

Figure 4.9

Electricity Prices in Short-Term Transactions in WEMI during August 2008 to September 2009

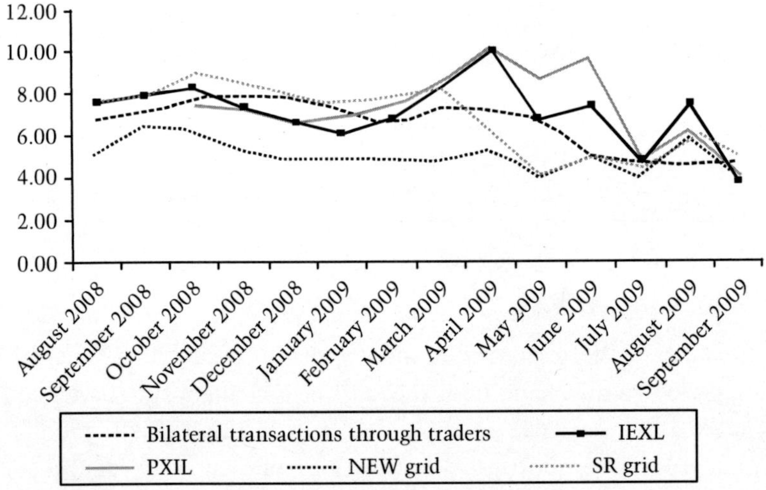

Note: Based on the data compiled from monthly report on short-term transactions of electricity available at the website of CERC, *www.cercind.gov.in* accessed in December 2009.

Source: CERC.

Table 4.8

Volume of Short-Term Transactions of Electricity in Power Exchanges during August 2008 to September 2009

Month	*Market Clearing Volume in Power Exchanges (IEXL and PXIL)*	*Short-Term Transactions*	*Volume in Power Exchanges to Short-Term Transactions (%)*
August-08	239.81	4831.74	4.96
September-08	278.54	4138.66	6.73
October-08	377.31	4232.18	8.92
November-08	470.05	4703.30	9.99
December-08	368.99	4833.12	7.63
January-09	316.43	4349.19	7.28
February-09	217.57	3935.62	5.53
March-09	428.61	4248.55	10.09
April-09	406.07	4432.07	9.16
May-09	341.70	4656.38	7.34
June-09	529.50	5065.64	10.45
July-09	495.16	5720.67	8.66
August-09	493.51	5789.29	8.52
September-09	527.23	5595.29	9.42

Note: Based on the data compiled from monthly report on short-term transactions of electricity available at the website of CERC, *www.cercind.gov.in* accessed in December 2009.

Source: CERC.

It may be seen from Table 4.9 that the weighted average prices in two power exchanges (IEXL and PXIL) show significant correlation of 0.84. The weighted average monthly prices of electricity in bilateral transactions through traders also show correlation with prices in power exchange (IEXL) and UI rates (SR grid), but no significant correlation is found between the prices in other mechanisms.

Table 4.9

Correlation of Weighted Average Monthly Prices of Electricity among Short-Term Transaction Mechanisms of Electricity during August 2008 to September 2009

Short-Term Transaction Mechanisms		*Bilateral*	*Power Exchanges*		*UI*	
			IEXL	*PXIL*	*NEW Grid*	*SR Grid*
Bilateral		1.00	0.54	0.42	0.16	0.76
Power exchanges	IEXL	0.54	1.00	0.84	0.46	0.47
	PXIL	0.42	0.84	1.00	0.04	0.15
UI	NEW Grid	0.16	0.46	0.04	1.00	0.49
	SR Grid	0.76	0.47	0.15	0.49	1.00

Similarly the weighted average daily prices in IEXL and PXIL show significant correlation of 0.69, but no significant correlation is found between the weighted average daily prices of electricity in power exchanges and UI rates as shown in Table 4.10.

Table 4.10

Correlation of Weighted Average Daily Prices of Electricity among Power Exchanges and UI Rates during August 2008 to September 2009

Short-Term Transaction Mechanisms		*Power Exchanges*		*UI*	
		IEXL	*PXIL*	*NEW Grid*	*SR Grid*
Power exchanges	IEXL	1.00	0.69	0.43	0.40
	PXIL	0.69	1.00	0.27	0.17
UI	NEW Grid	0.43	0.27	1.00	0.47
	SR Grid	0.40	0.17	0.47	1.00

Impact on the Bid Area Prices in Power Exchanges

As each electrical region of the country has been divided in two bid areas by the power exchanges, there are 10 bid areas for submission of bid prices. Based on the data of hourly bid area prices of electricity in IEXL and PXIL (available on their websites), it is observed that out of 11,040 hours during the period 28.6.2008 to 30.9.2009, the hourly bid area prices of electricity were not same in all the bid areas during 1,993 hours in IEXL. Similarly out of 8,208 hours during the period 24.10.2008 to 30.9.2009, the hourly bid area prices of electricity were not same in all the bid areas during 1,227 hours in PXIL. The highest area price difference was Rs 16.12/kWh in IEXL and Rs 8.38/kWh in PXIL. Figure 4.10 shows the difference in hourly bid area prices of electricity in both the power exchanges.

Figure 4.10

Difference in Hourly Bid Area Prices of Electricity in Two Power Exchanges (IEXL and PXIL) during the Period June 2008 to September 2009

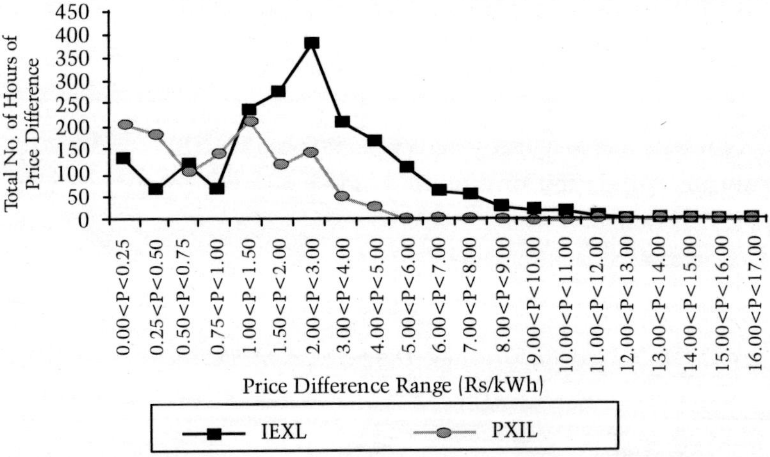

It may be observed from Figure 4.11 that in IEXL, there was no difference in hourly bid area prices of electricity till November 2008. During December 2008 to March 2009, difference in hourly bid area prices of electricity was noticed in some of the hours. The difference in hourly bid area prices of electricity was more prevalent during April-September 2009 and remained more than 50 per cent of the number of hours during June, July and August 2009.

Figure 4.11

Difference in Hourly Bid Area Prices of Electricity in IEXL during the Period June 2008 to September 2009

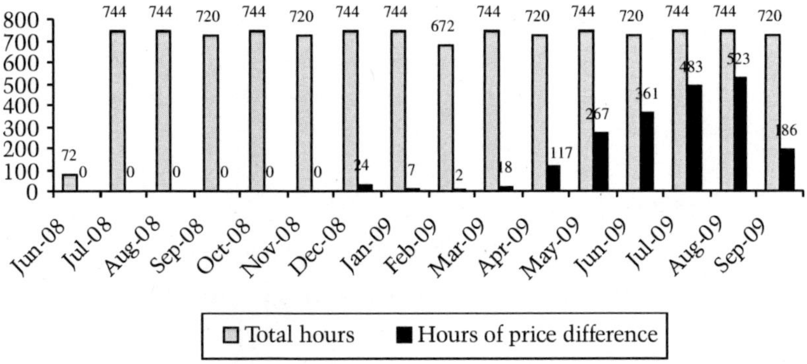

Similar trend is visible from the data of PXIL in Figure 4.12, which shows that there was no difference in the hourly bid area prices of electricity during October-December 2008, February 2009 and May 2009. During January, March and April 2009, the difference in hourly bid area prices of electricity was noticed in some of the hours. The difference in hourly bid area prices of electricity was more prevalent during June-September 2009 and remained more than 50 per cent of the number of hours during the months of July and August 2009.

Figure 4.12

Difference in Hourly Bid Area Prices of Electricity in PXIL during the Period October 2008 to September 2009

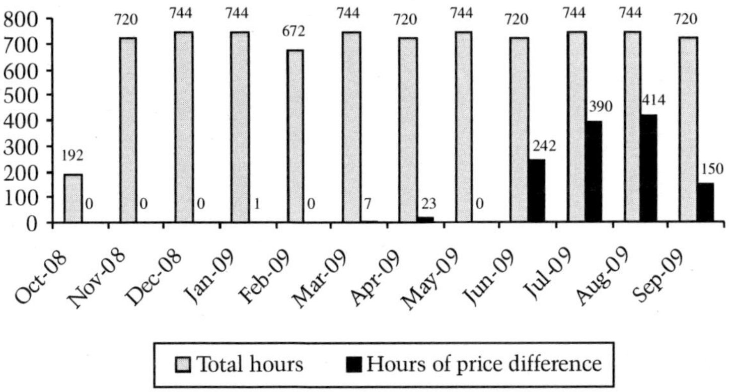

The difference in the hourly bid area prices of electricity in IEXL and PXIL indicate the constraints in transmission of electricity from one region to another region and more particularly transportation of electricity to northern region from other regions to meet the increased demand of electricity in summer months as the higher prices are observed on most of the instances in the northern region. This necessitates the need to create more transmission capacity for transfer of power from surplus regions to shortage regions.

Impact on Market Clearing Prices in Power Exchanges

The price difference between the two exchanges provides the arbitrage opportunities to the market participants. Such arbitrage opportunities exist normally in the initial phase of development of the power exchanges in view of the market inefficiencies. From the analysis of data during the period 24.10.2008 to 30.9.2009, it is observed that the difference in market clearing price between the two exchanges are on more frequent occasions and out of 8,208 hours during the period 24.10.2008 to 30.9.2009, the difference in market clearing price between IEXL and PXIL are found during

Figure 4.13

Difference in Hourly Market Clearing Prices of Electricity between Two Power Exchanges (IEXL and PXIL) during the Period October 2008 to September 2009

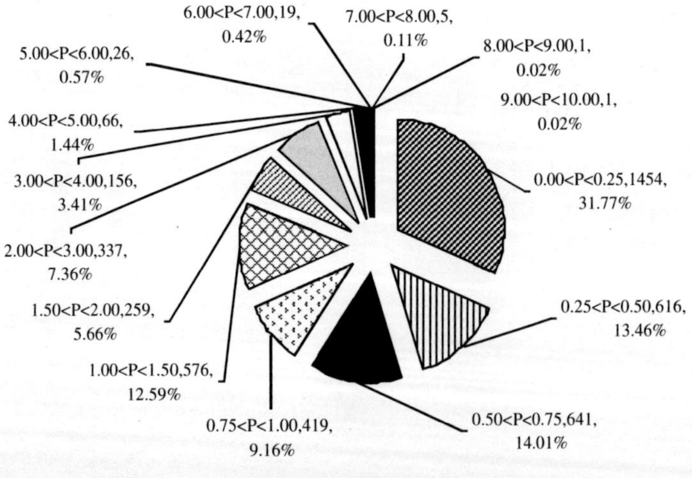

4,576 hours and the highest area price difference was Rs 9.84/kWh. Similar trend was also observed in hourly bid area prices of different regions between the two exchanges. Figure 4.13 shows the range of difference in hourly market clearing prices of electricity between two power exchanges (IEXL and PXIL).

From Figure 4.13, it may be observed that although the price difference was within Rs 1.00/kWh on 68.40 per cent of the instances, the price differences between Rs 1.00-1.50, Rs 1.50-2.00, Rs 2.00-3.00, Rs 3.00-4.00 and Rs 4.00-5.00 were also on substantial number of instances. While the price difference upto Rs 0.25 and Rs 0.50 may be ignored keeping in view the transaction cost, the higher price difference between the two exchanges indicates the inefficiencies in the market. Figure 4.14 shows the number of hours of difference in market clearing price of electricity between two power exchanges (IEXL and PXIL) during the period 24.10.2008 to 30.9.2009. It may be seen from Figure 4.14 that there was difference in the hourly bid area prices of electricity in all the months, but it was more prevalent during July, August and September 2009.

Figure 4.14

Number of Hours of Difference in Market Clearing Price of Electricity between Two Power Exchanges (IEXL and PXIL) during the period October 2008 to September 2009

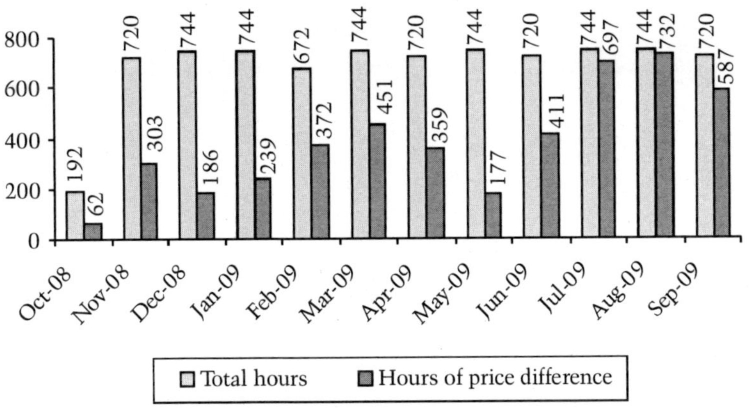

Conclusion and Suggestions

The basic objective of development of electricity market and start of power exchanges was to increase the competition and bring down the electricity prices in WEMI. However the results and findings of the study indicate that the electricity prices in all the mechanisms of short-term transactions of electricity have increased in WEMI after start of power exchanges. In the context of the findings of this study, following suggestions are made:

Adequate Transmission Capacity

The technology of electricity production and delivery make a reliable transmission network a necessary condition for an efficient wholesale market (Wolak, 2005b). The difference in the hourly bid area prices of electricity in power exchanges, IEXL and PXIL, indicates the constraints in insufficient transmission capacity of electricity from one region to another region and more particularly transportation of electricity to northern region from other regions. Even if there is sufficient transmission capacity, the congestion may occur due to blockage of transmission capacity by the market participants. The constraints in transmission capacity may give the market power to the regional markets. For example, when demand is very high in summer months in northern region and there typically does not exist enough transmission capacity to fully satisfy this demand with imports of relatively low-cost supplies from generators located elsewhere in the eastern and north eastern regions, northern region, at these times, becomes a 'load pocket' and generators located in northern region may have market power, when imports are constrained due to transmission constraints (Shukla and Thampy, 2009; Joskow and Tirole, 2000). As the market power of the electricity generator is a function of the size of the geographic market and location of final demand, a generator would possess local market power, as the transmission network does not provide the supplier with sufficient local competition to control its bidding behaviour in the wholesale market (Wolak, 2005a). The inter-regional transmission capacity in India is not sufficient for transfer of electricity between regions and transmission congestion[10] are experienced mostly in ER-NR and ER-WR links (CERC, 2006) as is evident from the measures initiated to reduce transmission congestion by imposing congestion charges[11] and to mitigate the market power by fixing the minimum and maximum price ceiling for inter-state day-ahead transactions in power exchanges and bilateral market. Based on the projected capacity addition during Eleventh Plan (2007-2012), there would be need to transfer the electricity between regions, particularly from eastern and north eastern regions to other regions. Although, there is sufficient transmission capacity[12] for transfer of power between the regions; however in some of the inter-regional lines, the capacity is very close to the requirements. As the redundancy in the transmission capacity reduces the market power, more inter-regional transmission capacity for transfer of power from surplus regions to shortage regions may be developed.

Fixing the Maximum Price Cap

The difference in the hourly bid area prices of electricity in IEXL and PXIL and market clearing price between IEXL and PXIL indicate the possibility of use of market power. The experience of restructuring in England and Wales have shown significant market power problem at the generation level (Green and Newbery, 1992; Joskow, 1997; Newbery, 1995;

10. Central Electricity Regulatory Commission (Measures to relieve congestion in real time operation) Regulations, 2009 defines the congestion as a situation where the demand for transmission capacity exceeds the available transmission capability.

11. Central Electricity Regulatory Commission (Measures to relieve congestion in real time operation) Regulations, 2009 defines the congestion charge as the supplementary charge kicked in on one or more regional entities in one or more regions for transmission of power from one region to another or from one state to another within a region when the deviations from the schedule cause the net drawl of power in the inter-regional or intra-regional transmission links to go beyond the total transfer capability limit.

12. The region-wise projected availability, demand and deficit/surplus in 2011-12 during winter, monsoon and summer seasons in peak and off-peak periods (Annexure A-4.4) shows that in 2011-12, there would not be much requirement of transfer of electricity during off-peak period between regions except northern region, where there would be shortages in winter off-peak and summer off-peak period. During monsoon peak period also, there would not be much requirement of transfer of power between regions except western region, where there would be shortage to the extent of 1,535 MW requiring the transfer of power from other regions. During winter and summer peak period, there would be shortages in northern, western and southern regions and surplus in eastern and north eastern regions, requiring the transfer of power from these regions to other regions.

Wolfram, 1996; Wolak and Patrick, 1997; von der Fehr and Harbord, 1993) and certain generating stations at strategic locations on the grid charged prices six times higher than those of other generators before the regulator imposed a price ceiling (Office of Electricity Regulation, 1992; Joskow, 1997). The price cap has been used in many efficient electricity markets in various countries such as Australia,[13] Singapore[14] and UK[15] to mitigate the market power. Various studies have also suggested some type of price regulation and use of price cap for mitigation of market power (Joskow, 1997). However a low price cap on the spot market not only reduces the incentives to sign forward contracts with gencos, which increases the likelihood of a unilateral market power problem in the spot market in the future, but also disincentivises the generators to make investments in technologies and new generation (Wolak, 2005b).

CERC has been using the price cap on UI rate and has fixed the overall maximum ceiling of Rs 7.35/kWh at the grid frequency of 49.20 Hz subject to a ceiling of Rs 4.08/ kWh for coal and lignite-fired stations and stations burning only administered price mechanism (APM) gas and with the provision of additional 40 per cent UI charge, when grid frequency is below 49.20 Hz. However, in some of the transactions through bilateral contracts and power exchanges, the electricity prices crossed the maximum UI ceiling rate (CERC, 2008). In view of steep increase in short-term power prices, CERC, in September 2009, fixed the minimum and maximum price ceiling of Rs 0.10/Kwh and Rs 8.00/Kwh respectively for inter-state day-ahead transactions in power exchanges and bilateral markets for a period of 45 days. In view of power shortage situation in India, there is a need to continue the maximum price ceiling to control the increase in electricity prices.

13. Australia, which is viewed as one of the most efficient electricity markets in the world by IEA, uses the price cap (Energy Regulatory and Market Development Forum, 2008).

14. After New Electricity Market of Singapore (NEMS) started on 1 January 2003, potential market power of the generation companies was checked by a mandatory price cap set for 100 per cent of the total demand in 2003 and the generation companies were not allowed to raise prices beyond the price cap (Chang, 2007).

15. In UK, when after expiry of 'vesting' contracts, generators started using market power by increasing the market price in the spot market, the market power problem was resolved by the regulator by forcing an agreement with the generators to set caps on the pool price (Hunt and Shuttleworth, 1997).

Reducing the Demand-Supply Gap

It has been found that market restructuring tends to lead to higher electricity prices in countries, where there are insufficient capacities (Xu, 2004) and it can be said that shortage of electricity is one of the main reason for increase of electricity prices. To bridge the demand-supply gap, there is a need to take steps not only to add new generation capacities and improve the efficiency of existing generation capacities, but also to reduce aggregate transmission and commercial (AT&C) losses and adopt demand-side management and energy conservation measures at the consumer end. Ultra mega power projects (UMPPs), with unique operating framework and project financing mechanism by allocation of risks amongst the participants, are helping quicker capacity addition in thermal generation from coal, and there is a need to extend UMPP and similar mechanisms in generation of electricity from other sources such as hydro, gas and wind (Shukla, 2009a and 2009b). It is suggested systematic implementation of distribution reforms by separating agricultural feeders, privatisation of distribution circles and using modern technologies e.g. smart cards to limit theft to reduce AT&C losses. The demand-side management may be implemented by mandating consumption standards, replacing incandescent lamps with CFL lamps, energy efficient standards for new constructions and real time metering for heavy users (McKinsey and Company, 2008).

Public Data Release

The different studies have shown the positive correlation between public release of data and well functioning of the wholesale electricity market. The studies have also shown that asymmetric information between market participants are very effective device for implementing less competitive market outcomes, through either unilateral or coordinated actions (Wolak, 2005a). The restrictive data release policy was found to be one of the reason of California electricity crisis and the two price spike periods in New Zealand market in June from September of 2001 and 2003 as it made impossible to undertake independent analysis (Wolak, 2005b). This necessitates the need for public release of all the data submitted to and produced by the market and system operators as the readily available data enables all market participants to become more sophisticated players in the wholesale market and protect themselves against potentially harmful market outcomes. Public availability of data also enables other entities besides the market monitor and regulator to perform analysis of market performance (Wolak, 1999). The difference in hourly bid area prices of electricity in IEXL and PXIL and difference in hourly market clearing prices of electricity between two power exchanges shows the market inefficiency and require more transparency in the operation of power exchanges. Based on the experience of other efficient electricity markets of other countries, CERC may lay down the policy of release of data with name of the market participant and specific generation unit associated with each bid.

Regulatory Governance

The economic rationale for regulation is usually to mitigate monopolistic pricing (Baldwin and McCrudden, 1987). It is now felt across the countries in the world that there is a need for effective regulation for efficient working of the market mechanism. USA is also revisiting the policy as evidenced from the presidential address of Barack Obama quoted as below:

> ...Nor is the question before us whether the market is a force for good or ill. Its power to generate wealth and expand freedom is unmatched, but this crisis has reminded us that without a watchful eye, the market can spin out of control—and that a nation cannot prosper long, when it favours only the prosperous..."

It is fact that a well-functioning wholesale electricity market requires a prospective market monitoring process. We have seen that the market may sometimes create the California crisis like situation even in countries like USA, where the peak and energy shortages do not prevail. The Indian electricity sector is, at present, facing acute peak and energy shortages and the dependence only on the market mechanism may not give the desired results. The objective of the Electricity Act, 2003 is not only to promote the competition, but also to protect the interest of consumers. The focus of the electricity regulatory commissions has been to bring the competition through market mechanism and allow determination of prices through market forces of demand and supply and as such adequate attention have not been given to protect the consumers' interest. For example, Section 24 of the Act imposes the duties on the distribution licensees to maintain uninterrupted supply of electricity to the consumers conforming to the standards regarding quality of electricity. The state electricity regulatory commissions should prescribe the performance standards for the Discoms so that Discoms may sell only surplus electricity after meeting the demand of their state. REBs should watch the grid frequency more carefully and RLDCs should manage the available open access capacity so that the blockage of transmission capacity by the market participant is avoided.

The authors would like to thank S.L. Rao, Chairman, ISEC and Former Chairman, CERC; Dr V.K. Garg, Chairman, JERC; Prof. Gopal Naik, Chairperson FPM, IIMB and Prof. Deepak Malghan, IIMB, for their valuable suggestions and insightful comments to improve the quality and content of the paper. This work was partially supported by the Centre for Public Policy visiting associateship of IIMB. The views expressed in this paper are those of the authors and do not necessarily reflect the views of the organisation/department of the authors.

Annexure A-4.1

Existing/Planned Inter-Regional Power Transfer Capacity (MW)

	March 2009	*Addition 2009-2012*	*March 2012*
Dehri-Sahupuri 220 kV S/c	130		130
Sasaram HVDC back-to-back	500		500
Muzaffarpur-Gorakhpur 400 kV D/c (with Series Cap+TCSC)	2000		2000
Patna-Balia 400kV D/c (Quad)	1600		1600
Biharshariff-Balia 400kV D/c(Quad)	1600		1600
Barh-Balia 400kV D/c (Quad) [Barh Tr. System]		1600	1600
Sasaram-Fatehpur 765kV S/c [DVC,NK, Maithon Tr. System]		2100	2100
Gaya-Balia 765kV S/c [DVC,NK, Maithon Tr. System]		2100	2100
Sasaram bypassing (additional capacity)	500		500
TOTAL EAST-NORTH	6330	5800	12130
Budhipadar-Korba 220 kV 3 ckts.	390		390
Rourkela-Raipur 400 kV D/c with series comp.+TCSC	1400		1400
Ranchi-Sipat 400 kV D/c with series comp.	1200		1200
Rourkela-Raipur 400 kV D/c II with series comp. [ER-WR strengthening]		1400	1400
Ranchi-WR Pooling Pt. 765kV S/c		2100	2100
TOTAL EAST-WEST	2990	3500	6490
Vindhyachal HVDC back-to-back	500		500
Auriya-Malanpur 220 KV D/c	260		260
Kota-Ujjain 220 KV D/c	260		260
Gwalior-Agra 765 kV S/c [Sasan Tr. System]	1100		1100
Gwalior-Agra 765 kV S/c II ckt [NR-WR inter-regional strengthening]	1100		1100
Zerda-Kankroli 400kV D/c [NR-WR inter-regional strengthening]	1000		1000
TOTAL WEST- NORTH	4220	0	4220
Gazuwaka HVDC back-to-back	1000		1000
Balimela-Upper Sileru 220kV S/c	130		130
Talcher-Kolar HVDC Bipole	2000		2000
Upgradation of Talcher-Kolar HVDC Bipole	500		500
TOTAL EAST-SOUTH	3630	0	3630
Chandrapur HVDC back-to-back	1000		1000
Kolhapur-Belgaum 220kV D/c	260		260
Barsur-L. Sileru 220kV HVDC Monopole	200		200
Ponda-Nagajhari 220kV D/c	260		260
South-West HVDC link		1000	1000
TOTAL WEST-SOUTH	1720	1000	2720
Malda-Bongaigaon 400 kV D/c	1000		1000
Birpara-Salakati 220kV D/c	260		260
Siliguri-Bongaigaon 400 kV D/c (Quad) [Palatana Tr. System]		1600	1600
TOTAL EAST-NORTH EAST	1260	1600	2860
NER Pooling point-Agra HVDC Bipole [Lower Subansiri Tr. System]		6000	6000
TOTAL NORTH EAST-NORTH	0	6000	6000
Various 132kV inter-regional links	600		600
TOTAL (CUMULATIVE)	20,750	17,900	38,650

Source: Ministry of Power, 2009.

Annexure A-4.2

IEXL Day-Ahead Market Timeline for Actual Delivery

Time	*Trading Day-Action*
10 AM – 12 PM	Participants submit bid to IEXL for purchase/sale of electricity on next day.
11 AM	NLDC communicates to IEXL the list of interfaces/control areas/ regional transmission system(s), on which unconstrained flows are required.
1 PM	IEXL communicates to NLDC the interchange on various interfaces/ control areas/regional transmission systems as intimated by NLDC.
2 PM	In case of congestion, NLDC communicates to IEXL the period of congestion and available limit for scheduling on respective interfaces/ control areas/transmission system(s).
3 PM	IEXL submits application for scheduling of collective transaction.
4 PM	NLDC sends the details (scheduling request of collective transaction) to all the concerned RLDCs.
5:30 PM	NLDC/RLDCs confirm the accepted schedule to IEXL.
6 PM	RLDC issues schedule.

Source: IEXL website accessed in December 2008.

Annexure A-4.3

Bid Areas in IEXL

Bid Area	*Region*	*States Covered under Bid Area*
N1	Northern	Jammu and Kashmir, Himachal Pradesh, Punjab, Chandigarh, Haryana
N2	Northern	Uttar Pradesh, Uttaranchal, Rajasthan, Delhi
E1	Eastern	West Bengal, Sikkim, Bihar, Jharkhand
E2	Eastern	Orissa
W1	Western	Madhya Pradesh, Chhattisgarh
W2	Western	Maharashtra, Gujarat, Goa, Daman and Diu-1, Daman and Diu-2, Dadra and Nagar Haveli, North Goa
S1	Southern	Andhra Pradesh, Karnataka, Pondicherry (Yanam), South Goa
S2	Southern	Tamil Nadu, Kerala, Pondicherry (Puducherry), Pondicherry (Karaikal), Pondicherry (Mahe)
A1	North Eastern	Tripura, Meghalaya, Manipur, Mizoram, Nagaland
A2	North Eastern	Assam, Arunachal Pradesh

Source: IEXL website accessed in December, 2008.

Annexure A-4.4

Season-wise Availability and Demand for 2011-12 (MW)

Regions	*Availability*	*Demand*	*Surplus/ Deficit*	*Perce-ntage*	*Availability*	*Demand*	*Surplus/ Deficit*	*Perce-ntage*
	WINTER OFF-PEAK				WINTER PEAK			
Northern	29782	32767	(2985)	-10%	36251	46810	(10559)	-29%
Western	42147	34041	8106	19%	45528	50275	(4747)	-10%
Southern	30031	26103	3928	13%	33676	35426	(1750)	-5%
Eastern	27657	10946	16711	60%	30449	15998	14451	47%
North Eastern	2638	1862	776	29%	4218	2660	1558	37%
Total	132255	105719	26536	20%	150122	151169	(1047)	-1%
	MONSOON OFF-PEAK				MONSOON PEAK			
Northern	45394	32767	12627	28%	45394	42129	3265	7%
Western	43828	36659	7169	16%	45598	47133	(1535)	-3%
Southern	32203	26103	6100	19%	34648	33561	1087	3%
Eastern	27899	10946	16953	61%	29389	14314	15075	51%
North Eastern	5658	1862	3796	67%	5658	2394	3264	58%
Total	154982	108337	46645	30%	160687	139531	21156	13%
	SUMMER OFF-PEAK				SUMMER PEAK			
Northern	38407	42129	(3722)	-10%	40564	45874	(5310)	-13%
Western	43758	36659	7099	16%	46334	52370	(6036)	-13%
Southern	30817	26103	4714	15%	34462	37290	(2828)	-8%
Eastern	27453	10946	16507	60%	30244	15998	14246	47%
North Eastern	4613	1862	2751	60%	5403	2660	2743	51%
Total	145048	117699	27349	19%	157007	154192	2815	2%

Source: Central Electricity Authority (2007).

References

Baldwin, Robert and C. McCrudden (1987). *Regulation and Public Law*. London: Oxford University Press.

Besant-Jones, J. and B. Tenenbaum (2001). "The California Power Crisis: Lessons for Developing Countries", *Energy & Mining Sector Board Discussion Paper Series*, Paper No.1

Bhushan, Bhanu (2005). *ABC of ABT: A Primer on Availability Tariff*. Gurgaon: Power Grid Corporation of India Limited.

Brennan, T.J., K. Palmer and S. Martinez (2002). "Implementing Electricity Restructuring", *Environmental and Resource Economics* 22: 99-132.

Central Electricity Authority (2007). *National Electricity Plan*. New Delhi: Central Board of Irrigation & Power. April.

———. (2009). *Power Scenario at a Glance*. April. *[www] http://cea.gov.in* (30 May, 2009).

Central Electricity Regulatory Commission (2006). *Staff Paper on Developing a Common Platform for Electricity Trading. [www] http://cercind.gov.in* (30 November, 2008).

———. (2008). *Staff Paper on Measures for Restraining the Prices of Electricity in Short-Term Sale/Trading. [www] http://cercind.gov.in* (30 November, 2008).

Chang, Youngho (2007). "The New Electricity Market of Singapore: Regulatory Framework, Market Power and Competition", *Energy Policy* 35: 403-412.

Energy Regulatory and Market Development Forum (2008). *Information Report, 2008*, Volume I. Asia-Pacific Partnership on Clean Development and Climate, Australian Government.

Green, Richard and David Newbery (1992). "Competition in the British Electricity Spot Market", *Journal of Political Economy* 100(5): 929-53.

Hunt, Sally and Graham Shuttleworth (1997). "Introduction: Economic and Technological Principles in Designing Power Markets", in Hunt and Shuttleworth (eds.), *Competition and Choice in Electricity*. West Sussex, England: John Wiley & Sons Ltd. pp. 1-10.

Joskow, Paul L. (1997). "Restructuring, Competition and Regulatory Reform in the U.S. Power Networks", The RAND *Journal of Economics* 31(3):450-487. (Autumn), Electricity sector; *The Journal of Economic Perspectives* 11(3):119-138 (Summer).

Joskow, Paul L. and Jean Tirole (2000). *Transmission Rights and Market Power on Electric*.

McKinsey and Company (2008). *Powering India: The Road to 2017*.

Ministry of Finance (2009). *Economic Survey, 2008-09. [www] http://www.nic.in*

Ministry of Power (2009). *Annual Report 2008-09. [www] http://powermin.nic.in* (30 December, 2009).

Newbery, David M. (1995). "Power Markets and Market Power", *Energy Journal* 16(3):39-66.

Office of Electricity Regulation (1992). *Report on Constrained-on Plant*. Birmingham, England.

Obama, Barack (2009). Inaugural Speech at Presidential Oath Ceremony. *[www] http://news.bbc.co.uk/1/hi/world/americas/obama_inauguration/7840646.stm* (12 February, 2009)

Shukla, Umesh Kumar (2009a). "Ultra Mega Power Projects", *Yojana* 53(7):38-41, July.

———. (2009b). "Ultra Mega Power Projects: A Solution to Achieve the Mission of 'Affordable Power to All' in India?", *Fourth International Conference on Public Policy and Management*, Bangalore, India, Indian Institute of Management. 9-12 August.

Shukla, Umesh Kumar and Ashok Thampy (2009). "Analysis of Competition and Market Power in Wholesale Electricity Market in India", *Centre for Public Policy CPP Monograph Series* Number 4. Bangalore, India, Indian Institute of Management

Von der Fehr, Nils-Henrik Morch and David Harbord (1993). "Spot Market Competition in the UK Electricity Industry", *Economic Journal* 103:531-46, May.

Wolak, Frank A. and Robert H. Patrick (1997). *The Impact of Market Rules and Market Structure on the Price Determination Process*. England and Wales Electricity Market, February.

Wolak, Frank A. (1999). "Market Design and Price Behaviour in Restructured Electricity Markets: An International Comparison", in Takatoshi Ito and Anne Krueger (eds.), *Competition Policy in the Asia Pacific Region*, EASE Volume 8. University of Chicago Press. pp.79-134.

———. (2005a). "Managing Unilateral Market Power in Electricity", *World Bank Policy Research Working Paper* 3691. September. *[www] http://worldbank.org* (30 November, 2008).

———. (2005b). "Lessons from International Experience with Electricity Market Monitoring", *World Bank Policy Research Working Paper* 3692. September. *[www] http://worldbank.org* (30 November, 2008).

Wolfram, C.D. (1996). "Strategic Bidding in a Multi-Unit Auction: An Empirical Analysis of Bids to Supply Electricity in England and Wales Electricity Market", *Mimeo*, Harvard Department of Economics, July.

Woo, C.K., D. Lloyd and A. Tishler (2003). "Electricity Market Reform Failures: UK, Norway, Alberta and California", *Energy Policy* 31:1103-1115.

Xu, Yi-Chong (2004). *Electricity Reform in China, India and Russia, The World Bank Template and Politics of Power*. Northampton, USA: Edward Elgar.

Section II

Nuclear

5 Nuclear Energy

S.L. Rao

Introduction

India has almost 500 million households that have no access to electricity and burn dried twigs, branches and leaves in little unventilated hutments, leading to high incidence of respiratory diseases and tuberculosis. Not only does India need to increase electricity usage for economic development, it needs to ensure that its people are able to use a benign form of energy. However, much of Indian electricity depends on the rapacious burning of coal and that has adverse environmental consequences especially in creating green house gases and climate change. While improving usage efficiencies, reducing wastage as in transmission and distribution (T&D) losses, etc., improving thermal efficiencies certainly have to be pursued. It is also critical that more power be generated. Coal, hydroelectric, gas, nuclear and renewables are the chief methods and all have to be used. Each of them, except coal and hydro, are more expensive. It is vital therefore that there is political consensus on giving leadership to Indian consumers so that they understand that electricity has to be paid for at least to cover costs.

The Indian consumer is accustomed to paying subsidised costs. This has resulted in massive losses to distribution utilities (estimated at Rs 60,000 crore in 2009 and Rs 70,000 crore in 2010). They are accentuated by regulators who deny actual expenses into tariff increases and instead keep them as regulatory assets, for later adjustment, which could takes years if at all, since it awaits a time when costs are lower and instead of reducing tariffs, the regulatory assets can be adjusted to keep the tariffs at the ruling levels.

In this context the giving of free power to agriculture, allowing theft of electricity, not tackling high technical losses in distribution, inefficiencies in generation and transmission leading to (high losses in transmission, distribution, in leakage due to technical losses, and in billing and collection, with collusion by electricity enterprise employees, must be stopped by regulators.

The independent regulatory framework for energy in India (direct and indirect) covers all forms of electricity generation, transmission, distribution, trading, and other aspects, oil and gas pipelines and separately, atomic energy. Electricity in India is a subject over which governments at both the Centre and the states have legislative power, though central legislation has precedence over state legislation. There is a Central Electricity Regulatory Commission (CERC) and about 24 or so State Electricity Regulatory Commissions (SERCs), including some Joint Commissions for some states and union territories.

Thus, this paper takes a close look at the prevailing regulatory environment for the energy sector in India, and then considers what needs to be done to facilitate the expansion of nuclear energy.

Regulatory Legislation

For nuclear energy, there are the Atomic Energy Commission (AEC), and the Atomic Energy Regulatory Board (AERB), the latter reporting to the former, and the former chaired by the Secretary, Atomic Energy, in the Government of India. They regulate the industry, set standards, inspect for safety, inspect equipment manufacturers, etc. There is thus no independent regulator. It is all part of the government system and wrapped in opacity and secrecy.

For downstream oil and gas there is the Petroleum and Natural Gas Regulatory Board (PNGRB). There are Appellate Tribunals for electricity and oil and gas; and for competition issues, the Competition Commission of India and the Competition Appellate Tribunal. The Competition Commission of India (CCI) is beginning to insist on its jurisdiction over the sectoral regulators in the case of competition, mergers and acquisitions (M&A), in electricity and oil and gas. This will no doubt be settled in the courts.

Some aspects of energy regulation remain under government departments: Director-General of Hydrocarbons (for upstream regulation);

tariffs for oil and gas; coal (though an independent regulator has been talked about for some years).

In regulatory legislation on electricity, the CERC regulates Central government-owned undertakings (NTPC, NHPC, NLC, Power Grid Corporation of India Ltd., REC but not NPC), determines their tariffs; inter-state transactions—trading, transmission, exchanges and spot and futures trading; adjudicates inter-state disputes; specifies Grid Code; and has determined advance tariffs for new projects.

The SERCs determine the tariffs of electricity generated within the state; approve the tariff paid for electricity bought from outside the state; approve purchases by the distribution company in relation to its estimated requirements for the period; approve the purchases up to the point where the demand forecasts are met by stacking the offers, with the least cost offers being at the bottom (merit order) and hence the first to be procured; also to approve open access and decide on a surcharge to be paid over the transmission tariff.

The Appellate Tribunal for Electricity will hear appeals on Orders of the ERCs.

Exclusion for Atomic Energy

Section 173 of the Electricity Act, 2003 states that, 'the Act shall not apply if it is inconsistent with any provisions of the Consumer Protection Act, 1986; Atomic Energy Act, 1962; or the Railways Act, 1989'. Hence regulation and determination of tariffs of atomic energy is with the AEC. This is an anomaly since electricity flowing through transmission and distribution lines regulated by the ERCs cannot be distinguished by type of electricity-hydro, thermal, renewable, etc. In practice however, the nuclear plants coordinate with the CERC and follow their decisions. With the entry of private nuclear plants, this anomaly must be rectified and all energy must be regulated by the ERCs. Private investment requires clarity, predictability and transparency in regulation, which the ERCs have now been giving. However, the electricity regulator has no access to cost data from nuclear energy generators.

Tariff determination of electricity has to reconcile two conflicting objectives: stimulating growth of the sector which requires encouraging investment, and safeguarding the interests of the consumer. Electricity tariffs are determined under the tariff policy of the GoI. 'Accelerated growth of the generation capacity sector is essential to meet the estimated growth in demand. Adequacy of generation is also essential for efficient functioning of power markets. At the same time, it is to be ensured that new capacity addition should deliver electricity at most efficient rates to protect the interests of consumers'.

Government policy requires that power procurement for future requirements should be through a transparent competitive bidding mechanism using the guidelines issued by the Central government. These guidelines provide for procurement of electricity separately for base load requirements and for peak load requirements. This would facilitate setting up of generation capacities specifically for meeting peak demand.

A two-part tariff structure is expected to be adopted for all long-term contracts for facilitating Merit Order despatch (which prioritises the cheapest electricity for despatch before despatching more expensive electricity). According to the National Electricity Policy, the Availability Based Tariff (ABT) was to be introduced at state level by April 2006 (very few have implemented it even in early 2011 though some have begun preparations for it). This framework would be extended to generating stations (including grid connected captive plants of capacities as determined by the SERC). The appropriate Commission may also introduce differential rates of fixed charges for peak and off-peak hours for better management of load.

Under the policy, power purchase agreements (PPAs) should ensure adequate and bankable payment security arrangements to the generating companies in case of persisting default in spite of the available payment security mechanisms. Since nuclear generation is in 'must run' plants, nuclear power is likely to be used for base load through long-term PPAs).

Accelerated growth of the generation capacity sector is essential to meet the estimated growth in demand. Adequacy of generation is also essential for efficient functioning of power markets. At the same time, it is to be ensured that new capacity addition should deliver electricity at most efficient rates to protect the interests of consumers. This policy stipulates the following for meeting these objectives.

Tariff Limitations by SERCs

Many state regulatory commissions have avoided legitimate tariff increases by not taking into account of genuine expenses and instead classifying them as 'regulatory assets', which will earn a return. The asset value would be recovered at the time of a future tariff increase. In many states even the return on these assets is accumulated in this way. The result is a serious shortfall in liquidity for the concerned utility, and a tax burden since these assets constitute accrued revenues and are accounted in profit calculations. With the imminent introduction of International Financial Reporting Standards (IFRS), these revenues must be reversed, causing a significant hit to the bottom line of the utilities.

This has happened with many state electricity boards where regulators have held back legitimate expenses as regulatory assets in order to avoid raising tariffs. In Delhi and Mumbai, private distributing companies by themselves or in partnerships with the state government, have over Rs 5,000 crore of such regulatory assets that have been due to them for many years. As a result these companies are constrained for funds for capital expenditures, are unable to declare dividends, and a culture of low tariffs is created which is making the sector unviable. The companies also have difficulty in raising working capital from banks as a consequence.

Different tariff determination methods are available to buyers (distribution enterprises) and generation companies.

a. Generating companies can enter into power purchase agreements with inter-state selling prices decided by CERC and state level purchase price to be paid by the distributing company decided by SERC so that the tariff is in the merit order.

b. Tariff-based competitive bidding: This is usually accompanied by special fiscal preferences given by the government. For domestic coal, there has been the allocation of captive coal mines from those available with the nationalised coal companies.

c. Merchant Power-Spot, medium-term (month) and soon futures contracts; also long-term purchase agreements. The idea here is that these tariffs are outside the regulators' purview. Private long-term purchases are possible. However the tariffs on such contracts might be lower than available on opportunistic spot or medium-term transactions.

d. Unscheduled Interchange (UI) surcharge is a penal charge enabling a commercial mechanism to stabilise frequency in the grid. It is also used as price by some unscrupulous state agencies that starve their customers to take advantage of these high penal charges and helping to stabilise the frequency.

Retail tariffs are determined by the SERCs on the basis of an Annual Revenue Requirement (ARR). Various norms are in force for operating norms, capital costs, return on equity, fuel escalation costs, etc. Today such information on atomic energy is not in the public domain.

Subsidies are specifically mentioned in Section 65 of the Electricity Act, 2003. The provision relates to reimbursement of subsidies to the distributing companies who are required to sell below costs to certain consumers. There is no provision regarding Central government reimbursing subsidies. If nuclear power is more expensive, there may be need for such reimbursement by Central government to enable it to be on the merit order for SERCs.

At the retail level, SERCs have developed the mechanism of cross-subsidies and meet the costs of selling below costs to some consumers by charging more from others. The calculation has to be done in a way that the costs are recovered, but the supplier does not make extra profits. Hence when retail consumers shift from one supplier to another under open access, the relevant cross-subsidy so far recovered from the better-off consumer, must be identified and reimbursed to the supplier of the below-cost consumers.

Costs of Nuclear Energy

There is no public knowledge about the actual capital and running costs of generating nuclear energy. Dr M.V. Ramana's writings (also corroborated elsewhere by Dr A. Gopalakrishnan, former chairman of the Atomic Energy Regulatory Board) in the *Economic and Political Weekly* suggest that it is significantly more expensive than thermal energy on both counts. (For example, see Ramana, 2007). The crux of his argument is that there are huge subsidies on heavy water supplied by the Department of

Atomic Energy (DAE) to the Nuclear Power Corporation (NPC). In addition the NPC has access to cheap or (free) financing from government. Further, capital costs are high as is operation and maintenance expenditure. The cost of the initial loading of heavy water is also at a low price or by leasing, and thus subsidised. This gives an effective subsidy of over 15 per cent of the initial capital cost of the reactor.

It is generally recognised that there is insufficient and incomplete data on nuclear energy costs in India. The electricity regulator has no jurisdiction over it and has no access to cost data as is available to him in the case of all other electricity. There is considerable secrecy and this comes in the way of any transparent tariff regulation.

Imported reactors from Europe or the USA are also likely to be even more expensive and said to be escalating at 15 per cent per annum. The Middle East political turmoil in 2011 has raised crude oil prices to $110 a barrel, and might go up further. It has raised other fuel prices as well, in sympathy! In addition the March 2011 earthquake in Japan and its effect on the Fukushima nuclear power plants there that came dangerously close to melting down have raised more concerns about safety. There is no doubt that safety regulation for equipment and for operations will become more stringent, adding further to costs. Nuclear energy can be expected to become much more expensive than traditional coal fired thermal generation.

Accident Liability

Then there is the issue of accident liability which will add costs for insurance. If the end result is that nuclear energy is much more expensive than other sources there will have to be a subsidy mechanism from government funds if the nuclear energy is not to fall off the electricity regulators' merit order.

India's Atomic Energy Act, 1962 says nothing about accident liability or compensation in the event of an accident. India is still not a party to the relevant international nuclear liability conventions (the IAEA's 1997 Amended Vienna Convention and 1997 Convention on Supplementary Compensation for Nuclear Damage – CSC). At present and until the laws are changed to allow private entry, all civil nuclear facilities are owned and must be majority-owned by the Central government (companies like Nuclear Power Corporation of India Ltd. (NPCIL), only public sector enterprises). The liability issues arising from these installations are the government's responsibility. On 10 September 2008 the government assured the USA that India 'shall take all steps necessary to adhere to the Convention on Supplementary Compensation (CSC)' which was signed two years later. Under existing Indian legislation, foreign suppliers may face unlimited liability, which prevents them from taking insurance cover, though contracts for Kudankulam Nuclear Power Plant 1&2 exclude this supplier liability.

Both houses of Parliament have passed a bill related to third party liability. This was debated in the context of strong national awareness of the Bhopal disaster in 1984, probably the world's worst industrial accident. A Union Carbide (51 per cent US-owned) chemical plant in the central Madhya Pradesh state released a deadly mix of methyl isocyanate and other gases due to operator error and poor plant design, killing some 15,000 people and badly affecting some 100,000 others. The company paid out some US $1 billion in compensation—widely considered inadequate.

The new bill places responsibility for any nuclear accident with the operator, as is standard internationally, and limits total liability to 300 million SDR (about US $450 million) 'or such higher amount that the Central Government may specify by notification'. Operator liability is capped at Rs 1,500 crore (about US $330 million) or such higher amount that the Central government may notify, beyond which the Central government is liable.

However, after compensation has been paid by the operator (or its insurers), the bill allows the operator to have legal recourse to the supplier for up to 80 years after the plant starts up if the 'nuclear incident has resulted as a consequence of an act of the supplier or his employee, which includes supply of equipment or material with patent or latent defects of (or?) sub-standard services.' This clause giving recourse to the supplier for an operational plant is contrary to international conventions.

At the same time it is reported that negotiations with Russia for additional nuclear reactors at Kudankulam are held up because of this sub-clause, in this case involving Atomstroyexport (Russia). The original

Kudankulam agreement said that supplier liability ended with delivery of the plant. US diplomatic sources are similarly opposed to supplier liability after delivery.

The bill also does not make any mention of India ratifying the Convention on Supplementary Compensation for Nuclear Damage (CSC), or any other international treaty or framework governing nuclear liability under which the supplier cannot be sued in their home country. The CSC is not yet in force internationally, but Indian ratification would bring it closer to being so, and was part of the September 2008 agreement with USA. In October 2010 India signed the CSC.

In October 2010 it was reported that NPCIL proposed to set up a fund of Rs 1500 crore ($336 million) for nuclear liability 'with the Centre addressing anything over this level'. This legislation imposes obstructions to foreign entry as well as any Indian private entry into nuclear energy when the policy is changed to allow them to enter.

Regulation of Public-Private Partnerships (PPPs)

If private investment is to enter by itself or jointly with public enterprises into nuclear generation, any public private partnerships (PPPs) structures must fit in to the regulatory regime of the Nuclear Suppliers Group (NSG), AERB, IAEA, as well as the Indian Electricity Act 2003, the Indian Contract Act, 1860 and the Atomic Energy Act, 1962. If there are additional costs in conforming to these, they will also have to come into the calculations. If there are joint ventures between government-owned and private companies, many regulatory issues must be resolved, starting with corporate governance rules under the Indian Companies Act, 1956.

Financing Nuclear Energy

The ambitious targets for nuclear energy generation will call for mobilising substantial financial resources within and outside India, and will need the backing of the Indian government, as well as the Indian and foreign private sector and foreign governments. For this to happen, all conflicting issues must be amicably resolved. These include liability and pricing. Given the growth story in India and the market for power, financing nuclear energy may not be difficult. However, the sector must be made financially viable by ensuring that consumers pay appropriate tariffs and all enterprises make a reasonable return on investment. This is not the case today where the state electricity boards have a loss in 2010 of Rs 60,000 crore and many private distributing enterprises have large amounts blocked as regulatory assets. Government must intervene and ensure that regulators do not prevent tariff increases in relation to costs.

Land and Environment Concerns

There are also environmental and land compensation issues to be resolved. These will persist at any location. Environmentalists and non-governmental organisations (NGOs) are now fully aware of the potential compensation and how to agitate for it. The struggle in Jaitapur where Areva SA (France) is to put up 10,000 MW of nuclear energy capacity is the first of such agitations in the case of nuclear energy. After the accidents at Fukuyama in Japan, accentuated by the cluster of six reactors in one place, it is unlikely that public opinion will accept 10,000 MW (6 reactors of 1,660 each) of nuclear generation capacity at one location, even though the location is in an area that is in seismic zone 3 and located at a height so that it is very unlikely to face tsumami waves. If such concentration is not there, costs will be higher per plant. The draft bill before Parliament on land acquisition might make it easier to acquire land but it will certainly make it more expensive. In any case, we can expect considerable debate and greater clarity on environmental and safety issues in future and much greater community consultation.

In strategic terms India's entering into agreements on nuclear energy generation with a variety of governments (Russia, France, Canada, USA, South Korea, etc.), signals a major change in direction since India's independence. India is now firmly lined up with the developed world. This will have many beneficial results but also make India a target for China and Pakistan. We can expect a competitive race from Pakistan in building nuclear power plants as well as offensive weapons and their delivery systems. We can expect a great deal of aggression from China on our borders. At the same time we will have the support of other major world economies. India is poised to become a major military, energy and economic power. This will bring responsibilities as well.

Issues to be Addressed

There are huge costs to nuclear generation and many environmental issues as well as land acquisition that must be dealt with. Many legislative changes will need to be undertaken to enable private and foreign participation in nuclear generation. Human resources must be developed. Indian capacity to make plant and equipment must be developed. The following are the major issues that must be addressed:

i Who will regulate tariffs?

ii Costs disclosure: so far it is not in the public domain. Will it be so for private sector or private-public partnerships?

iii. If nuclear energy costs are high, will there be subsidies to enable it to be in the merit order? Who will pay and when? What is the certainty of payment?

iv. Will atomic power plants come under the jurisdiction of the ERCs?

v. Long-term power purchase agreements are vital for 'must run' generation plants as are nuclear power plants. Will these have to be approved by Government of India (GoI) or will there be a policy directive from GoI instructing the regulator to accept the PPA as approved by government?

vi. How will a two-part tariff structure and a competitive bidding process work for atomic energy? Plants would have been set up on a public-private partnership or invitation basis and costs may prevent the tariff from being competitive.

vii. The electricity distribution enterprises in India sell much of their electricity below cost and are accumulating huge loss. (Estimated at around Rs 60,000 crore in 2010.) Political leaders and regulators prevent tariff increases to match costs. This has to change if more expensive nuclear energy is to enter the system.

viii. Fast reactors and thorium reactors are the key to energy security and their development and deployment has to be core long-term objective. PHWRs, FBRs and thorium reactors are linked through respective fuel cycles and would need sequential development.

ix. Strong emphasis on human resource development to continue.

References

Ramana, M.V. (2007). "Heavy Subsidies in Heavy Water: Economics of Nuclear power in India", *Economic and Political Weekly,* August 17.

Section III

Renewable Energy

6 The Case for Wind Energy in India

Gaurav Gandhi, Pradeep Kumar and Chintan Shah

Introduction

Indian government is now visibly keen to tap renewable energy (RE) source, amply available in the county to support huge electricity quench, born out of tremendous infrastructure and industrial development. The government is also in the process of unbundling the electricity sector to make RE technologies more comparable to conventional power and also evolve more maturely among other RE technologies. Central Electricity Regulatory Commission (CERC), inline to the Government's vision, has come up with tariff orders, appropriately supporting RE development in the country balancing investment interests, electricity needs and climate issues. This paper discusses different electricity policy and regulatory initiatives taken by Government of India (GoI) to deploy renewable energy into the national electricity grid. These initiatives demonstrate multi-pronged, long-term and integrated strategies for achieving key goals in the context of climate change. It deliberates on the evolution of Ministry of New and Renewable Energy (MNRE) from Commission for Additional Sources of Energy (CASE). The paper explores different developments that have taken place by way of regulatory and policy interventions to bolster renewable energy, subsequent to the Electricity Act, 2003, in time to address climate change issues at national and sub-national level. It has discussed various financial incentives such as Generation Based Incentives (GBI), technical up-gradation such as Indian Electricity Grid Code 2010 (CERC, 2010), Introduction of the ambitious and revolutionary Renewable Energy Certificates (RECs) and the National Action Plan on Climate Change (NAPCC). This paper also discusses different drivers for trend change and its impact on technology and in commercial terms. The paper concludes on commercial impacts, quality improvements and further ingenuity needs from GoI on policy and regulatory abutting current developments, highlighting the probable strives to achieve targets set under NAPCC and the commercial opportunities for further investments in this sector.

Institutional Development

The Ministry of New and Renewable Energy (MNRE) is the nodal ministry of the Government of India for all matters relating to new and renewable energy. Evolution of MNRE started with the formation of Commission for Additional Sources of Energy (CASE) in 1981 which was further rechristened to Department of Non-Conventional Energy Sources (DNES) in 1982 undertaking additional responsibilities of formulating policies and their implementation, programmes for development of new and renewable energy apart from coordinating and intensifying R&D in the sector. It was further renamed to Ministry of Non-Conventional Energy Sources (MNES) in 1992 finally coined to Ministry of New and Renewable Energy (MNRE) in October 2006. Under MNRE, three specialised technical institutions were formed for solar energy, wind energy and bioenergy. Technical needs in the area of solar energy are addressed by Solar Energy Centre (SEC), Centre for Wind Energy Technology (C-WET) for wind energy and, Sardar Swaran Singh National Institute of Renewable Energy (SSS-NIRE) for bioenergy. Apart from technical institutions, MNRE has instituted a special financial institution, more popularly known as the Indian Renewable Energy Development Agency Ltd. (IREDA). It is a non-banking financial institution providing term loans for renewable energy and energy efficiency projects.

Post-Electricity Act (2003) Environment

The Electricity Act, 2003 (EA 2003) played a pivotal role for unbundling of the electricity sector and discussed in detail on renewable energy, for the first time in its history. So much so that it has a separate mention on renewable energy.

The Electricity Act, 2003

This act abolished all other earlier electricity related laws,[1] i.e., (i) The Indian Electricity Act, 1910, (ii) The Electricity (Supply) Act, 1948 and (iii) The Electricity Regulatory Commissions Act, 1998. Salient features of EA 2003 are as follows:

- Special section for renewable energy, i.e., Section 86(1)e.
- Provision of Renewable Portfolio Standard (RPS).
- Captive generation has been freely permitted.
- No licence required for generation and distribution in notified rural areas.
- Provision for private licensees in transmission.
- Trading, a distinct activity monitored by Regulatory Commissions.
- Open access in distribution.
- For rural and remote areas stand alone systems for generation and distribution permitted.

Impact of EA 2003 on Wind Power Development

Installation of renewable energy witnessed massive growth after enforcement of EA 2003, especially in the wind energy sector. Wind energy market in India is mature in terms of technology while other renewable sources are still in the nascent phases of development. Also, deployment of wind turbines is comparatively easy which aided the upsurge in wind energy installation in India. Currently it represents India as fifth largest wind power installer after Spain, with 75 per cent share of domestic renewable installation. This can be well indicated with Figure 6.1, which shows yearly installation of wind energy in India. It clearly shows that year 2003 was defining year for RE deployment.

Incentives and Subsidies

Ministry of New and Renewable Energy (MNRE) introduced certain guidelines on renewable power purchase in 1993. As per these guidelines, states were allowed to purchase renewable electricity at Rs 2.25/kWh, with

1. The Electricity Act, 2003.

Figure 6.1

YoY Wind Power Installation in India and Different Policy Milestones

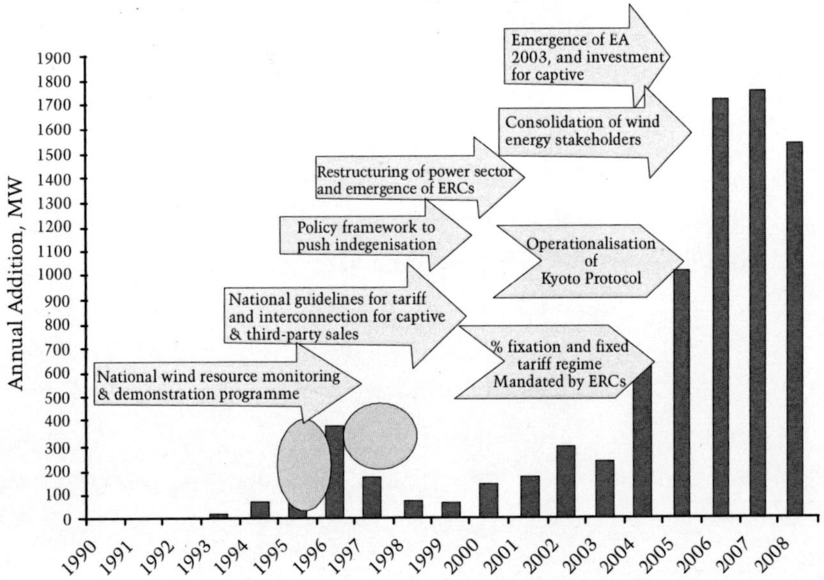

an annual escalation of 5 per cent. Also certain wheeling and banking charges were incorporated in these guidelines. But, these were just guidelines and none of the states were bound to purchase renewable electricity at that time. These guidelines were adopted with different variations by concerned utilities in the states. Actual mandate on purchase of renewable electricity came into force after the implementation of EA 2003. The concept of 'feed' in tariff was introduced and different states offered different electricity purchase tariff, rationally aligned with available wind resources.

Direct Fiscal Incentives

As on date, GoI offers various direct fiscal incentives to promote renewable energy deployment for both grid connected and off-grid applications. These fiscal incentives have proven to be pivotal in attracting project developers in the renewable energy sector.

Accelerated Depreciation (AD)

Under Section 32 of Income Tax Act (IT), project developer (investor) can avail up to 80 per cent depreciation in the first year.[2] In addition to 80 per cent depreciation, investor can also claim 20 per cent depreciation, for the projects after 2005, for new plant and machinery, mainly as a tool to attract new technology to optimise wind resource of the country. This proved an attractive tool by which most of the profit making corporate diverted their tax liabilities to renewable sector.

Tax Holiday

Section 80 (I) A of IT affirms that income generated from renewable sources by sale of electricity will be exempted from tax liabilities for 10 years,[3] and this can be flexibly adjusted over the project life, but in a continuous manner, i.e., across 10 years in the entire length of the project.

Subsidy

Certain renewable energy components are subsidised in terms of custom duty. Also, some of the renewable products such as aero-generator, water pumping wind mills and battery chargers are exempted from excise duty.

Contribution of Renewables in India

Looking at the latest trends, it can be safely said that the renewable energy sector is one of the most prominent sectors under the capacity addition programme. If we look at the progress of total installed capacity against actual planning under different five year plans, it will be noticed that there is a deviation of almost 50 per cent in planning and actual capacity addition, which is shown in Table 6.1.

---

2. Vipradas, M. (2008). Renewable Energy in India.
3. Ibid.

Table 6.1

Five-Year Plan for Installation

Five-Year Plan	*Total Installed Capacity (MW)*		
	Projected	*Actual*	*%*
8th	30540	16420	54
9th	40250	19015	47
10th	41110	19010	46

But contrary to the above, if we look at RE capacity addition during the Tenth Five-Year Plan, it will be noticed that the total RE installed is twice than its original planned capacity addition. This clearly shows the maturity level in the field of technology and ease of its installations. RE installation in Tenth Five-Year Plan is shown as under:

Table 6.2

RE Capacity Addition in Tenth Five-Year Plan

	Total RE Installed Capacity: 10th Five Year Plan (MW)		
	Projected	*Actual*	*%*
Wind	2200	5426	246
Other RE	1184	1369	115
Total RE	3384	6795	201

Out of 18,450 MW installed capacity of renewable energy installed in India, wind power installations stand at 14,126 MW, which constitutes for more than 75 per cent of total RE installed in the country (as on March 2011). Further, it is worthwhile to note that in India, RE installations account for around 11 per cent of the total electrical installed capacity and contribute to around 6 per cent of electricity generation.

Various Off-take Arrangements for Wind Power Projects

There are four broad categories under which a developer can register his project.

Captive Utilisation

Wind turbines can be used for captive electricity consumption. This off-take arrangement is attractive to large corporate having large electricity consumption. After paying wheeling/transmission charges investor gets a set-off in their monthly electricity bills. Some states are also offering banking facility for captive consumption. This can be illustrated with Figure 6.2.

Figure 6.2

Captive Utilisation

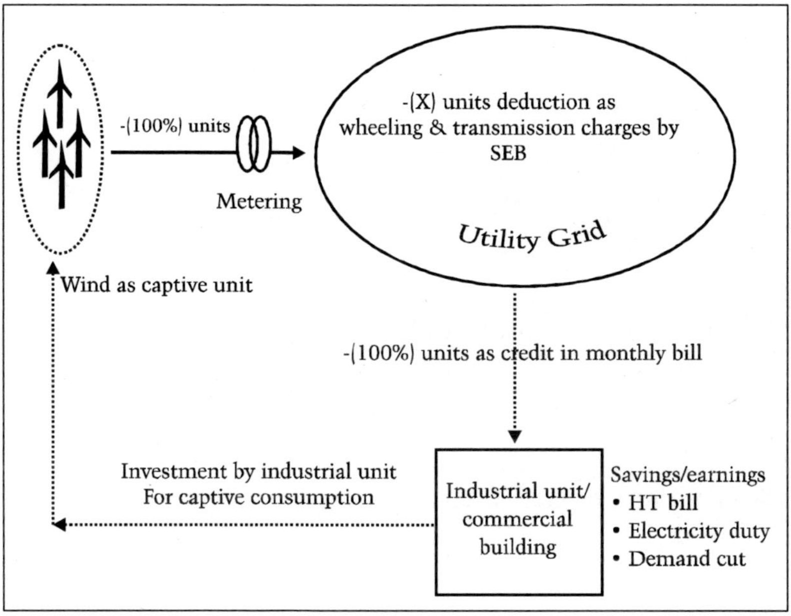

Procurement by Discoms-Feed in Tariff

With provision of Renewable Purchase Obligation (RPO) in EA 2003, it is now mandatory for states to procure certain percentage of their electricity from renewables. Respective state commissions specify feed in tariffs (FIT) and concerned Discoms purchase the renewable electricity from generator at specified FIT in that state.

Figure 6.3

Discom Procurement

OA/Merchant Generation

Most of the states in India allow open access sale of electricity. In this category, investor has option to sell electricity to an open access consumer after paying open access and transmission charges defined by respective state regulatory commissions.

Figure 6.4

Open Access Generation

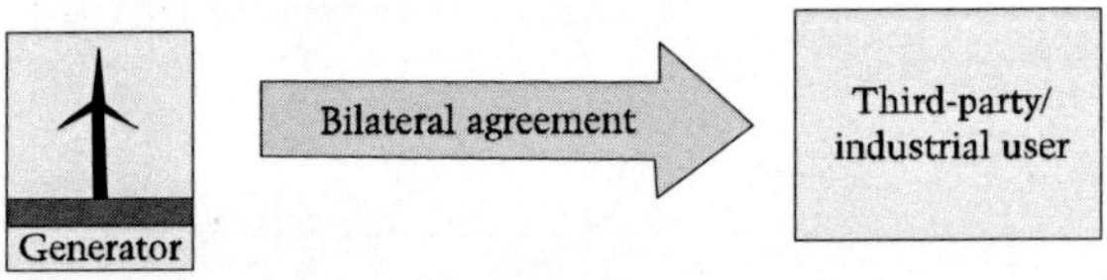

Renewable Energy Certificate (REC)

This is comparatively a new concept in India. A renewable electricity generator is eligible to get renewable energy certificates for its generation of non-electricity attribute, which can be traded on power exchanges and price discovery for REC is on trading basis. The minimum price that a non-solar REC can fetch is Rs 1.5/kWh to maximum Rs 3.9/kWh.

The electricity attribute of this renewable electricity generation can be procured by Discoms at annual power purchase cost (APPC) price for earlier year, notified by concerned SERCs.

Figure 6.5

REC Arrangement

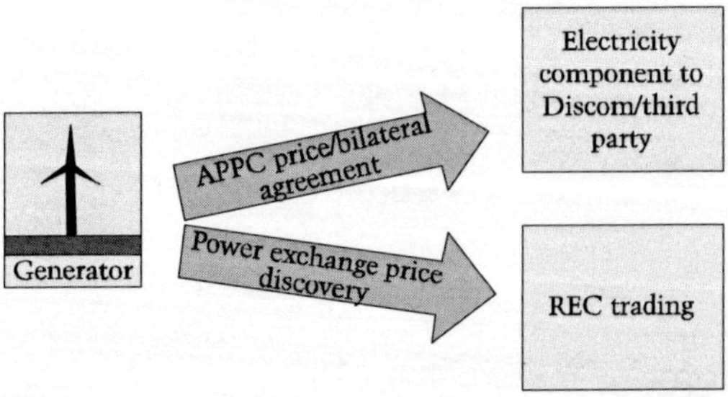

A detailed study of REC mechanism is discussed in later pages.

Recent Developments

Recently GoI has come up with radical policy announcements, which clearly envisages its vision of addressing climate issues and seriousness in deploying renewable energy. GoI has proactively initiated process to support renewable energy deployment through appropriate channels. GoI has released National Action Plan on Climate Change (NAPCC) and Central Electricity Regulatory Commission (CERC) provided guidelines to deploy different renewable sources with the state governments adopting them. CERC also announced proposed enforcement of Renewable Energy Certificates (REC) mechanism along with MNRE announcing the Generation Based Incentive (GBI) scheme. All these efforts demonstrate a national level commitment for a sustainable clean environment by contributing higher renewable electricity mix to the national grid. Different initiatives by GoI to promote renewable energy deployment, issued recently, are discussed in detail below.

NAPCC

GoI released NAPCC on 30 June 2008,[4] to outline its strategy to meet the challenge of climate change. It outlines a national strategy that aims to enable the country adapt to climate change and enhances the ecological sustainability of India's development path. It prescribes 5 per cent of renewable electricity into grid by 2010 and 1 per cent annual increase till 2020, i.e., by 2020 total renewable electricity into grid totalling 15 per cent. Eight National Missions, form the core of the NAPCC. These Missions are as below,

1. National Solar Mission.
2. National Mission on Enhanced Energy Efficiency.
3. National Mission on Sustainable Habitat.
4. National Water Mission.
5. National Mission for Sustaining the Himalayan Ecosystem.
6. National Mission for a Green India.
7. National Mission for Sustainable Agriculture.
8. National Mission on Strategic Knowledge for Climate Change.

4. National Action Plan on Climate Change, 2008.

Consequent to the announcement of the NAPCC, which identified development of solar energy technologies in the country as a National Mission, the GoI has approved Jawaharlal Nehru National Solar Mission (JNNSM)[5] which aims at development and deployment of solar energy technologies in the country by 2022. It aims to deploy solar technologies in three phases, spanning the remaining period of the Eleventh Plan and first year of the Twelfth Plan (up to 2012-13) as Phase 1, the remaining four years of the Twelfth Plan (2013-2017) as Phase 2 and the Thirteenth Plan (2017-2022) as Phase 3.

Table 6.3

Target under JNNSM

	Unit	*Target for Phase I*	*Cumulative Target for Phase II*	*Cumulative Target for Phase III*
Grid connected (33 KV and above) solar plants	MW	1000	4000	20000
Off-grid solar applications	MW	200	1000	2000
Solar collectors	Million Sq. m	7	15	20

Policy Initiatives to Achieve Targets under JNNSM

- o National Tariff Policy, 2006 would be modified to mandate that the SERCs fix a percentage for purchase of solar power, i.e., separate provision of Solar-RPS within the existing RPS framework. The Solar-RPS for states may start with 0.25 per cent in Phase 1 and to go up to 3 per cent by 2022.
- o Twenty-five year (long term) power purchase agreements (PPAs), to start with under Phase 1, to be signed between NTPC Vidyut Vyapar Nigam Ltd. (NVVN) and Solar Power Developers.
- o Exemptions on specific capital equipment, critical materials, components and project imports.

Generation Based Incentive (GBI)

Due to provision of accelerated depreciation, many independent power producers (IPPs), NGOs, trusts, academic and research institutions faced

5. MNRE (2010). Jawaharlal Nehru National Solar Mission.

difficulties since they were not able to utilise the depreciation benefit. Hence to attract IPP investments and to enhance quality of energy generation, MNRE proposed GBI structure which was notified in December 2009.[6]

Eligibility for GBI

- o The GBI scheme would be applicable only for those power producers who do not avail of the accelerated/enhanced depreciation benefits under the Income Tax Act.
- o The scheme will be applicable only for those IPPs having minimum installed capacity of 5 MW and whose capacities are commissioned for sale of power to the grid after the announcement of the scheme.
- o The scheme will not be applicable to those who have set up capacities for captive consumption, third-party sale, merchant plants etc.

The interface between MNRE and investor is IREDA. All financial transactions will be carried out through IREDA. The scheme is as below:

- o GBI will be provided to the wind power producers at Rs 0.5/kWh of power fed in the grid for a period of not less than four years and a maximum period of 10 years.
- o The maximum incentive available to a wind power project will be Rs 6.2 million per MW and the total disbursement in a year will not exceed one fourth of the maximum limit of the incentive (i.e., Rs 1.55 million per MW during the first four years).
- o The scheme is to be applicable to a maximum capacity of upto 4,000 MW during the remaining period of Eleventh Plan period. However, the provision of accelerated depreciation is to be continued till the Eleventh Plan period or introduction of Direct Tax Code, whichever is earlier. GBI will be applicable to wind turbines commissioned after the issue of this scheme and commissioned on or before March 31, 2012.

CERC

Central Electricity Regulatory Commission (CERC) is regulatory body formed under EA 2003 to advocate states on different issues related to

6. MNRE (n.d.). Generation based Incentive, 2009.

electricity pertaining to tariff determination, inter-state electricity transmission regulation, adjudication upon disputes involving generating companies or transmission licensee, maintaining grid standards etc.

CERC Tariff Guidelines

CERC has notified renewable energy tariff benchmarking guidelines, which are well accepted by SERCs. The guidelines call for tariff setting for renewable energy projects to be based on resource availability, e.g. for wind energy, CERC has classified four different zones, with reference to wind power density of the region and differential feed in tariff is estimated for respective zones.

Table 6.4

CERC Tariff

WPD (Watts/m2)	*PLF/CUF (Per cent)*	*Levelised Tariff as per CERC Regulations & Format Prescribed (Rs/kWh)*
200-250	20	5.60-5.70
250-300	23	4.90-5.00
300-400	27	4.15-4.25
>400	30	3.60-3.65

These guidelines will support RE deployment by minimising risk of resource variation and will attract investment in RE projects.

CERC (Terms and Conditions for Recognition and Issuance of Renewable Energy Certificate for Renewable Energy Generation) Regulations, 2010

In January 2010, CERC has notified REC mechanism.[7] The certificates will be dealt through power exchanges only, as approved by CERC. Need for REC mechanism was recognised against the background of uneven spread of renewable energy resource in India with most of the blessed states having good renewable resource, and associated RPO being met. Additionally, the aim is to incentivise renewable electricity production, to distribute renewable electricity prices evenly and to increase public participation in the process. Due to higher cost associated with solar

7. Central Electricity Regulatory Commission. *http://www.cercind.gov.in/*

power projects, RECs are further categorised as solar and non-solar and will be traded separately in respective sections. CERC approved auditors will monitor these REC transactions.

Working of REC Mechanism

Every State Electricity Regulatory Commission (SERC) appoints an entity known as state agency, responsible for accreditation of renewable energy generator. Renewable energy is generated and injected in the grid by accredited RE generator. The energy which is fed in the grid by the renewable energy generator is consumed by the load connected to the grid. The quantum of energy so injected is metered and accounted by State Load Despatch Center (SLDC). The injected energy is certified by SLDC/state agency and is registered with Central agency which is National Load Despatch Center (NLDC).

RE generator after getting registered with Central agency will apply for issuance of REC based on its certified energy. After verification, REC Registry issues the REC to RE generator. One REC will be equivalent to 1 MWh of renewable energy certified. RE generator after getting the REC will sell it in power exchange. Power exchange then verifies with REC Registry before settlement of the trade, selling the certificate.

Obligated entity *viz.*, the distribution companies, captive users and the open access users will buys these RECs from power exchange for fulfilling their renewable purchase obligation which they could not do through buying of renewable energy directly from RE generator. REC is extinguished from the account of RE generator in REC Registry as its gets traded at power exchange. Obligated entity surrenders REC purchased in order to meet its renewable purchase obligation. There is a compliance auditor appointed by CERC that verifies the compliance.

Figure 6.6

REC Mechanism

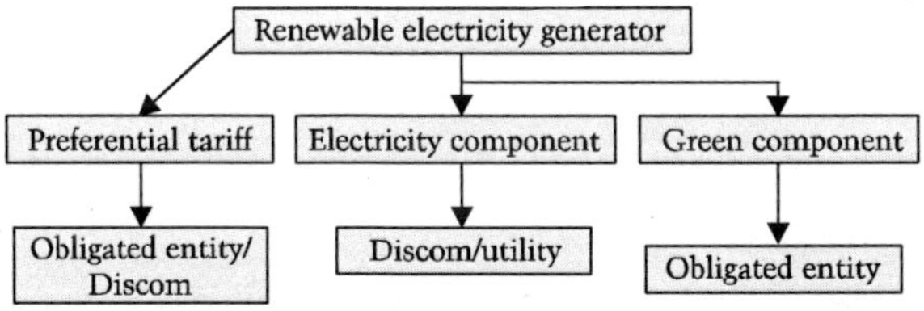

CERC has also notified certain forbearance and floor prices for solar and non-solar RECs as below:

Table 6.5

Forbearance and Floor Price for REC Mechanism

	Non-Solar REC (Rs/MWh)	*Solar REC (Rs/MWh)*
Forbearance price	3,900	17,000
Floor price	1,500	12,000

As CERC has notified REC framework, some SERCs have already started to adopt it. Once SERC notify, REC mechanism will be in operation, which is expected to be by this year end.

Indian Electricity Grid Code 2010

CERC notified The Indian Electricity Grid Code 2010 ('IEGC') in April 2010.[8] It prescribes rules, guidelines and standards to be followed by various participants in the system to plan, develop, maintain and operate the power system in the most secure and efficient manner, while facilitating healthy competition in the generation and supply of electricity.

IEGC has a separate section which provides guidelines on scheduling of renewable energy generators connected to the grid. For renewables, this code will be applicable from January 2011 and for non-renewables it is already in force since May 2010. Once it is enforced for renewable, scheduling of renewable electricity is must and within tolerable limits stated in IEGC. Failing which there are commercial implications by way of Unscheduled Interchange (UI) charges.

The Trend Change

The change of trend can be categorised into technology and commercial trend change.

Technology

With such massive techno-policy revamps and new players exploring Indian market, shift in technology is eminent. Even, historically,

8. CERC. Indian Electricity Grid Code, 2010.

technology became more sophisticated over the time. Earlier machines were stall controlled, controls located in nacelle, more electromechanical operation oriented, but now most of the machines are advanced pitch controlled and have inbuilt controllers for safety purposes. Capacity of the turbine has also increased over the time. Increase in capacity deployment over time is shown in Figure 6.7 and increase in capacity factor due to improved technology is shown in Figure 6.8.

Figure 6.7

Installation Trend

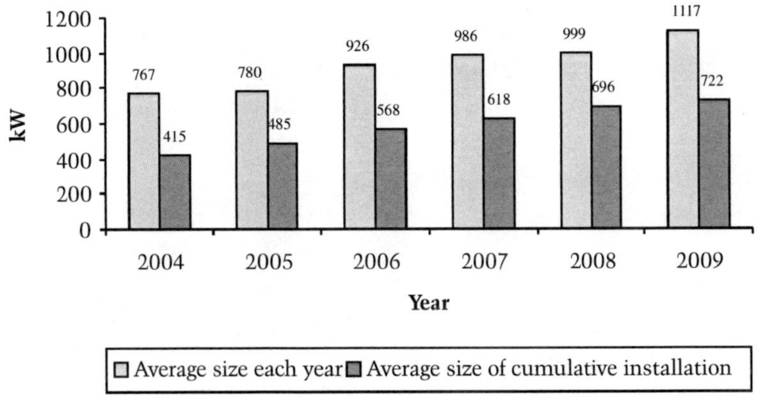

Figure 6.8

CUF Trend

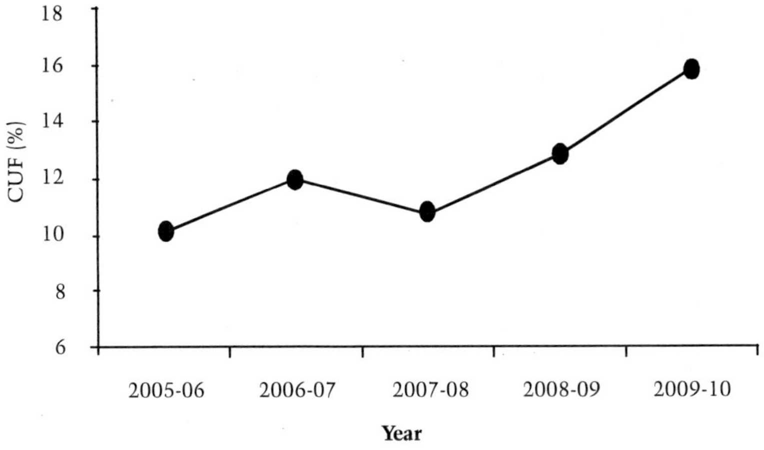

Advanced power electronics and condition monitoring softwares are integral part of the system involving supervisory control and data acquisition (SCADA) connectivity that detects any uneven operational malfunction (controllers are programmed accordingly), and reacts aptly, reducing machine downtime and enhancing design life of machine. Reliability of the components is amplified by means of appropriate national and international test standards such as IEC, DEWI etc. Emphasis on power conditioning is increasing and machines are incorporating power conditioning systems such as LVRT, harmonics controller etc. In the near future, the industry is also set to follow IEC standards on harmonics. Even in terms of towers, the earlier erected lattice type structures are now replaced by tubular towers. Concrete type towers for bigger machines are also introduced which are more cost-effective than the tubular towers and provide equal strength. Generators of machines are also being upgraded over the time. Currently most of the manufacturers provide synchronus and asynchronus machines, while many of the new entrants are offering permanent magnet synchronous generator (PMG) and double fed induction generator (DFIG) technology. Many turbine manufacturers are also offering elevator for the machines having higher hub heights, which further reduces time to restore machine for generation. Some manufacturers are offering machines with transformer yard at nacelle which helps to reduce transmission losses. Several optimised blade designs with elegant aerodynamic capabilities are commercially available. Improved aerodynamics is resulting into higher energy yield and lower dynamic stresses. For extreme weather conditions such as in Rajasthan, appropriate cooling measures are deployed within the machines.

All these developments in technology and design are a result of broader operating experience of machines in different states and different operating conditions.

Commercial

Apart from domestic corporate investors, with introduction of GBI and REC, Indian wind energy sector is also open and lucrative to the untapped segments such as IPPs, NGOs, trusts etc. The minimum revenue, through sale of power at pooled cost and sale of RECs, can be expected to be around Rs 3.7/kWh and maximum of Rs 7.5/kWh. The average tariff one can expect is around Rs 6.1/kWh, which is significant compared to existing feed in tariff (FIT).

Table 6.6

REC Scenario

All the Price Assumptions are on Rs/kWh Basis	*Min REC (at Floor Price)*	*Avg. REC*	*Max REC (at Forbearance Price)*	*Assumptions*
With min. pooled cost	3.7	4.9	6.1	Min. pooled cost=1.68
With max. pooled cost	4.6	5.8	7.0	Max. pooled cost=2.57
With CDM and max. pooled cost	4.7	6.1	7.5	
Assumptions	Min. CDM=0.15	Avg. CDM=0.325	Max. CDM=0.5	

Conclusion

GoI has proactively participated to revamp the electricity sector by appropriately imposing policy and regulatory reforms. To achieve target of 15 per cent renewable by 2020 specified under NAPCC, GoI has announced JNNSM programme ensuring solar energy contribution. Also, solar energy is promoted under REC mechanism by segregating it as Solar REC and defining higher commercial attributes. IEGC 2010 has ramped up quality standards of electricity grid by addressing unevenness in interconnection and electricity despatch. GBI mechanism is announced commensurately to tap IPPs, NGOs, academic and research institutions, trusts etc., distinctly different than typical depreciation based investors, which will also help the sector to evolve qualitatively and more professionally, meeting the electricity goals.

Though, GoI has taken such serious steps to amplify renewable energy use, more endeavors are necessary in policy and regulatory terms. The good reason for saying this is—projected electricity demand statistics. As per GoI published data, electricity requirement of country by 2021-22 will be 1,914 billion units (BUs).[9] As per NAPCC target considering 15 per cent of electricity from various renewable technologies, projected renewable electricity generation stood around 287 BUs, assuming 28 per cent

9. 17th Electric Power Survey of India, 2009.

cumulative capacity factor for all renewable energy technologies (in consideration with technological advances), there is requirement of 117,081 MW of renewable energy to be installed by 2021-22. Of this, renewable energy installation by March 2011, stands around 18,450 MW. That means the required installations are around 100,000 MW till 2021-2022, which is 8,340 MW per annum! Now even if we consider that solar installation is in line to the targets as per JNNSM, there is requirement of around 6,500 MW non-solar renewable to be deployed to the National grid annually, starting from year 2010-11. As GoI has already set installation targets for solar, it is now required to set installation targets for non-solar renewable more ambitiously in view of plausible mature renewable technology available immediately, such as wind energy. On the medium-term, GoI cannot afford the risk, in terms of time and investment, by waiting to induct another mature renewable energy technology. Hence, there is inevitable dependency on wind power sector. There is immediate requirement to triple the annual wind power installation from 2010-11, which was 1566 MW in 2009-10. To maintain corporate interest and attract IPPs, it is imperative that the accelerated depreciation benefits should continue in parallel with GBI benefit, distancing business interests. There is also an urgent need to adopt low carbon economy by affiliating to energy efficiency techniques and prioritising RE technologies, such as solar lantern, solar street lighting etc., wherever possible. Further, the recently introduced cess on coal should be diverted to further subsidising renewable electricity. Also, there is an urgent need to address and redefine R&D for technology enhancement, which will also reflect in higher energy yield, utilising available resource in a more optimised fashion.

References

Indian Electricity Act, 1910.

The Electricity (Supply) Act, 1948.

The Electricity Regulatory Commission Act, 1998.

7 Renewable Energy Certificate Mechanism

Balawant Joshi

Introduction to Renewable Energy Certificate (REC) Mechanism

On June 30, 2008, the Hon. Prime Minister of India announced the National Action Plan for Climate Change (NAPCC) which envisages several measures to address the problems of global warming. The Plan emphasises the priority of maintaining a high economic growth rate and identifying measures that promote developmental objectives while yielding the benefits of solving climate change effectively. The NAPCC has identified that increasing the share of renewable energy in total electricity consumption in the country is one of the important measures to combat climate change. NAPCC has set the Dynamic Minimum Renewable Purchase Standard (DMRPS) of 5 per cent renewable energy purchase for the FY 2009-10 against the current level of around 4 per cent. Further, the NAPCC envisages that this target will increase incrementally by 1 per cent over the next 10 years, which would constitute approximately 15 per cent of the portion of renewable energy generation in the total energy mix of India by 2020. This would require a quantum jump in the deployment of renewable energy across the country. Strong policy measures, proactive regulatory frameworks, and innovative financing instruments would be required, if the desired level of penetration of renewable energy is to be achieved. One such policy instrument prescribed in the NAPCC is the Renewable Energy Certificate (REC) mechanism, which would enable a large number of stakeholders to purchase renewable energy in a cost-effective manner.

Taking into consideration the recommendation of the NAPCC for the promotion of renewable energy, the Ministry of New and Renewable Energy

(MNRE) had engaged ABPS Infrastructure Advisory Private Limited (ABPS Infra) for conducting a study to develop a 'Conceptual Framework for the Proposed REC Mechanism in India'. The report submitted by ABPS Infra included its recommendations on the operational and institutional framework for the REC mechanism in India. Along with the MNRE, the Central Electricity Regulatory Commission (CERC) and the Forum of Regulators (FOR) have also been engaged in the conceptualisation process.

Regulatory Initiatives for the REC Mechanism

In order to accelerate the large scale deployment of renewable energy, the CERC, on January 14, 2010, had notified its Regulations 2010, wherein the terms and conditions for recognition and issuance of RECs for renewable energy generation are outlined. Implementation of these regulations would enable obligated entities to meet their RPO targets through RECs obtained from entities in other states which may have generated the RE. An expert committee has been formed to enable successful implementation of the REC mechanism in India.

The CERC, in its regulations, has designated the National Load Despatch Center (NLDC) as the Central agency that would take up such functions as, registration of RE generators as eligible entities to receive the REC, issuance of RECs, maintaining and settling accounts in respect of RECs, serving as the repository of transactions in RECs, etc. The regulations notified by the CERC and the FOR on the REC framework also envisages development of various processes and procedures; from accreditation of the RE generator to redemption of the RE Certificate. In order to operationalise the REC mechanism in India, ABPS Infra assisted the NLDC in preparation of the procedures which would provide guidance to the various stakeholders and also play a crucial role in the successful implementation of the REC mechanism.

The CERC regulations also envisages that the price of the REC would be determined through the market mechanism on the exchange platform. However, in order to safeguard the interests of various stakeholders, it is also envisaged that the price of the REC would be governed by the forbearance (ceiling) price and the floor price, which shall be determined by the CERC from time to time. In order to operationalise the REC mechanism, the CERC has recently issued an order in the matter of determination of forbearance and floor price for the REC framework.

In addition to the initiative taken by the CERC, the Forum of Regulators (FOR) has approved of the model regulations on the Renewable Purchase Obligation (RPO), and compliance and implementation of the REC framework for the State Electricity Regulatory Commission (SERC). Essentially, the regulations specify that fulfillment of RPOs can be met through purchase of RECs. They also specify the roles and responsibilities of various entities, involved at the state level, in implementation of the REC mechanism.

Key Drivers for the REC Mechanism in India

The REC mechanism essentially seeks to address the mismatch between availability of RE sources and the requirement of the obligated entities to meet their RPO across states. So far, inter-state exchanges of RE were constrained by the fact that such transactions were governed by inter-state open access regulations and the regional energy accounting framework, which necessitated scheduling of power. Some of the RE sources, such as biomass power or bagasse based cogeneration can be scheduled and inter-state open access transactions effected, based on whether they have been scheduled in the past. However, inter-state exchanges of power, based on non-firm RE sources, such as wind energy, solar power, small hydro power, etc., was constrained. Besides, the cost of open access wheeling under the long-term arrangement was prohibitive for such non-firm RE sources due to their inherent lower capacity utilisation factors. The REC mechanism addresses these constraining factors as the certificate is issued for the energy generated at the point of injection into the grid.

Thus, it is envisaged that the REC mechanism would facilitate the emergence of a large number of cross-border RE transactions based on non-firm RE sources, while at the same time, enhancing the volume of cross-border RE transactions based on firm RE sources as well.

While effective implementation of inter-state transactions would be the primary objective of the REC mechanism in India, some other objectives identified are:

- Reducing costs for RE transactions.
- Increasing flexibility for participants to choose the participant in RE transactions.
- Overcoming geographical constraints to harness available RE sources.
- Effective implementing of RPO regulations in all Indian states.
- Creating competition among different RE technologies.
- Developing an all encompassing incentive mechanism.
- Reducing risks for local distribution licensees.

Concept of the REC Mechanism in India

While, conceptually, the Indian operational framework is similar to REC mechanisms implemented in other parts of the world, it has been customised to comply with India's existing legal and regulatory framework. As per the REC mechanism in India, the purchase of an REC shall be deemed as the purchase of power generated from RE sources. Thus, the RE generator will be selling two independent and exclusive products from the same quantum of RE generation. These products will be: 1) electricity and 2) the electricity's associated environmental attributes in the form of RE certificates.

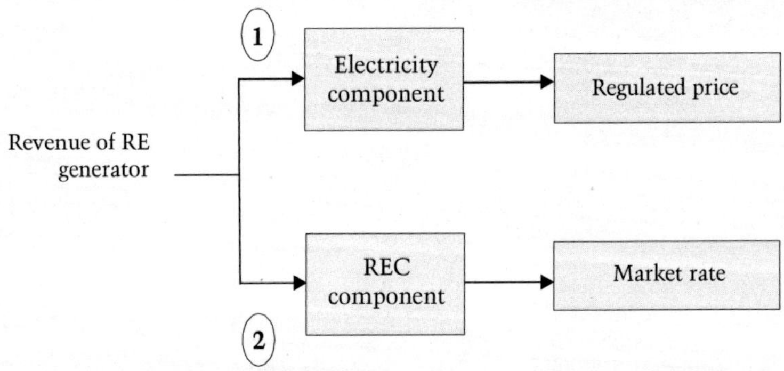

Under the REC mechanism, one REC will be issued to the RE generator for each MWh of electrical energy fed into the grid. The RE generator may sell electricity to the distribution company at its weighted average pooled cost of power purchase, approved by the concerned SERC for the previous financial year, and its RECs to any buyer, at the market price through the exchange mechanism in a transparent manner. The RE generator may sell the certificates, only through the power exchange, to those entities who have to meet with their RPO target as well as to voluntary buyers. The purchase of RECs will be deemed as purchase of power generated from renewable sources and, accordingly, will be allowed for compliance with the RPO target. The REC mechanism will enable obligated entities in a state to procure RECs generated from any of the states in India and surrender the same to satisfy its RPO target.

Operational Framework for the REC Mechanism in India

The schematic depicted in the Figure 7.2 represents a flow diagram for various processes involved in the operation of the REC mechanism. The numbers indicate the chronological sequence of seven identified key processes.

The operational framework for the REC mechanism may be sub-divided into four categories, namely:

a. Accreditation of the renewable energy generation project with the state agency.
b. Registration of the renewable energy generation project as an eligible entity with the Central agency.
c. Issuance of the renewable energy certificate to the eligible entities by the Central agency.
d. Purchase and redemption of the renewable energy certificates by the obligated entities or voluntary buyers.

The operational framework of the REC mechanism is elaborated below.

Step 1 and 2: Accreditation and Registration of the Renewable Energy Generator

The state agency, as may be designated by the concerned SERC, shall undertake the accreditation of the renewable energy generator; and the entity, which has received a 'certificate of accreditation' from the state

Figure 7.1

Schematic of Operational Framework for REC Mechanism

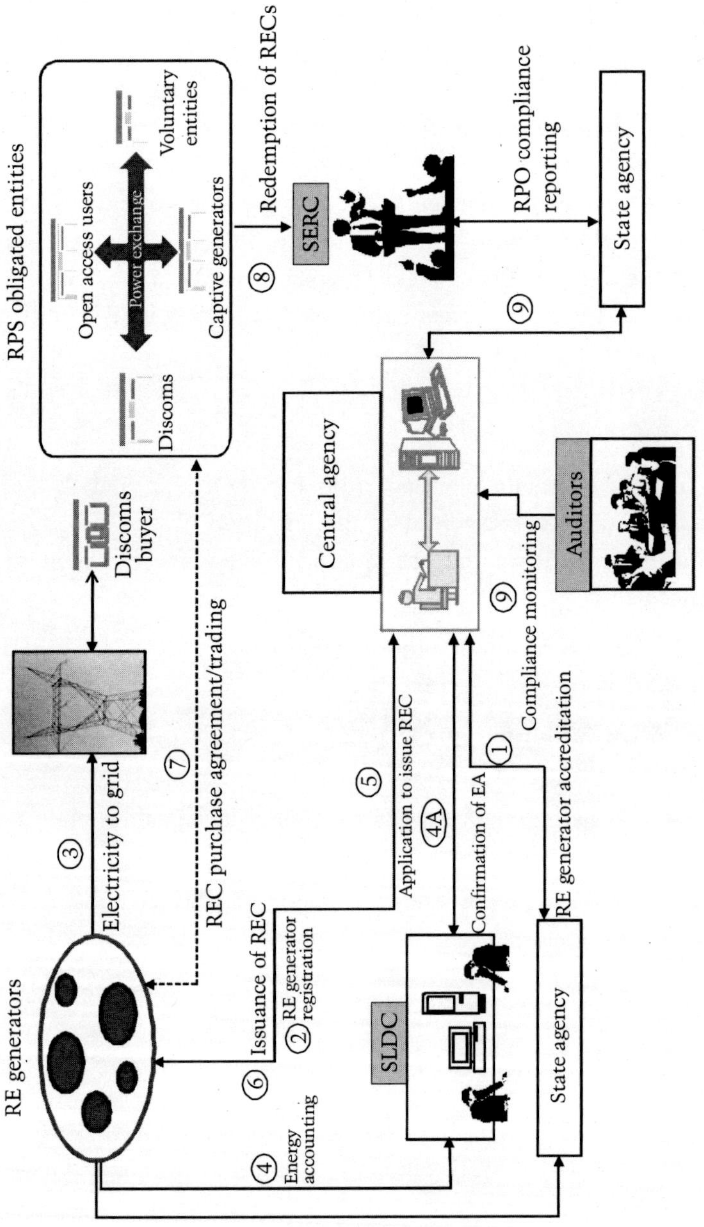

agency, shall apply to the Central agency for its registration as an eligible entity for receiving the renewable energy certificate.

Step 3: Electricity Generation and Feeding into the Grid

The electricity generated in an RE project is injected into the grid. This electricity is consumed in real time by load prevalent in the system, which in turn is accounted against the consumption by the entities which had a contract with that particular RE project.

Step 4: Energy Accounting by the State Load Despatch Centre

The metering of the quantum of electricity injected into the grid by the renewable energy generator and energy accounting shall be carried out by the concerned State Load Despatch Centre (SLDC). The SLDC shall provide the energy injection report to the renewable energy generator and shall also submit the report for the energy accounts of RE projects to the Central agency, as per established procedures, on a regular basis. The Central agency, while issuing the renewable energy certificates to the renewable energy generator, shall take into account this energy injection report prepared by the SLDC.

Step 5: Application for Issuance of REC

The RE generator will apply to the REC Issuance Registry (i.e., Central Agency) to issue the RE certificates equivalent to the amount of electricity injected into the grid and as certified by the SLDC.

Step 6: Creation and Issuance of RECs

Referring to the generation report submitted by the SLDC and state agency, the REC Registry will create and issue the appropriate number of RECs to the concerned RE generator (the 'eligible entity').

Step 7: Sale of the REC by the RE Generator (Eligible Entity)

Once the RECs are issued to the RE generator (eligible entity), the sale/purchase of RECs amongst various RE generators, obligated entities and voluntary buyers is proposed to be undertaken only through the power exchange operational under the guidance of the CERC.

Step 8: Surrender/Redeeming of RECs

The obligated entities can procure the RECs over the exchange platform and need to surrender the RECs to the SERCs to meet their RPS obligation. Thus, a convenient and effective mechanism for ensuring RPO compliance by the obligated entities is effectuated. The REC Registry shall maintain the record of RECs issued and RECs received for redemption on a regular basis.

Step 9: Compliance Reporting

It is envisaged that the Central Commission, in consultation with the National level registry (i.e. Central agency), may appoint from time to time compliance auditors to inquire into and report on the compliance of REC regulations by the persons applying for registration, or on the compliance by the renewable energy generators in regard to the eligibility of those entities being issued certificates and all matters connected thereto.

Summary

For a nation-wide implementation of the REC framework, it is essential that uniform policies are adopted at the state level in consonance with the regulatory framework formulated by the CERC and the FOR. Till date, very few SERCs have issued the draft/final regulations on the RPO; its compliance and implementation for the REC framework for their respective states, in line with the model regulations specified by the FOR.

Another important element that would ensure the success of the REC scheme would be the commitment, at the state level, to fix RPO targets, keeping in view the current national target of 5 per cent with the increasing trajectory, as envisaged in the National Action Plan on Climate Change (NAPCC), so that there is sufficient demand for RECs; which in turn would fuel growth of RE capacity addition irrespective of location. Besides, limitations under the present RPO regulations in many states, such as the lack of an adequate enforcement mechanism and the scope of 'obligated entities' to include open access and captive consumption, also needs to be addressed. Among other challenges in the implementation of the REC mechanism, there would be the challenge of capacity building of the various stakeholders and agencies, *viz.*, the Central agency, the state agencies, SLDCs, and other stakeholders, including the renewable energy generator and the obligated entities.

Implementation of the REC mechanism would essentially overcome geographical constraints to harness available RE sources and would also aim to reduce costs associated with RE transactions. It would provide increased flexibility for obligated entities to fulfill their renewable purchase obligation and thus pave the way for a healthy, competitive environment among different RE technologies.

For further details, visit *http://www.abpsinfra.com* and contact at *balawant.joshi@abpsinfra.com*

Section IV

Coal and Gas

8 The Need for a Robust and Transparent Regulatory and Policy Framework for Growth of the Gas Sector in India

Krishna Sarma

Natural gas is fast emerging as the preferred fuel of the future, in view of it being a clean and economically attractive fuel and also a desirable feedstock. The *Hydrocarbon Vision 2025* reiterates that increased focus needs to be given to the natural gas sector.

Indian consumption of gas has risen faster than any other fuel in recent years. According to a recent FICCI-Mckinsey paper (2010), India's natural gas demand was expected to nearly double to 320 million metric standard cubic metre per day (mmscmd) by 2015 from the current demand of 166 mmscmd—made up of nearly 132 mmscmd supplies from domestic fields and the rest from imported liquefied natural gas (LNG). According to this report, the demand would be driven by the refining and petrochemical industry (35 mmscmd), the power sector (5 mmscmd) and city gas distribution (about 10-12 mmscmd). Further, there was a total potential of 310-320 mmscmd, if gas is used to ease peaking power deficits. It is important to note here, that in the near past there has been substantial downward revision from previous forecasts because of the volatility of crude prices internationally; problems with securing reasonably priced LNG sources, and financing LNG import projects; financial problems in the power sector; political and diplomatic challenges in ensuring overland transnational pipelines from Iran, Bangladesh and Myanmar; and so on.

India's policy in the oil and gas sector from 1970 through till the late 90s was geared towards encouraging the hegemony of the public sector undertakings. Till recently, gas production in India was the preserve of the national oil companies—Oil and Natural Gas Corporation Limited

(ONGC) and Oil India Limited (OIL). Though incorporated, every aspect of the businesses of these companies was under the control of the Ministry of Petroleum & Natural Gas (MOP&NG). In 1984, the Gas Authority of India Limited (GAIL) was formed to establish a gas pipeline grid and to manage the retailing of natural gas.

After the opening of the oil and gas sector, there are Indian and foreign private companies—stand alone or through joint ventures (JVs)—which are engaged in upstream and downstream oil and gas activities. Since 1997, the Government of India, under the New Exploration Licensing Policy (NELP) has given concessions to private and public sector companies through production sharing contracts (PSCs) with the right to market gas at market determined prices.

Currently, more than 4,000 kms of pipeline infrastructure has been laid and marketing of a substantial portion of the gas is done by GAIL. To promote and develop the gas pipeline network and city or local gas distribution networks, the Government has established a Gas Advisory Body. The Dahej-Vijaipur (DVPL) and the Hazira-Vijaipur-Jagdishpur Gas (HVJ) pipelines that serve the states of Maharashtra, Gujarat, Madhya Pradesh, Uttar Pradesh and Delhi was decided keeping in mind the considerations of political economy. The reforms now permit and encourage private participation and 100 per cent FDI through the automatic route of laying of gas pipelines and facilitating open access for all players to the pipeline network on a non-discriminatory basis. The list of proposed pipelines include, the Dabhol-Bangalore gas pipeline passing through Maharashtra and Karnataka; and the Kochi-Kanjirkkod-Bangalore/ Mangalore pipeline passing through Kerala, Tamil Nadu and Karnataka. Except in Rajasthan, OIL markets its own gas, while Cairn Energy markets its gas from the Lakshmi fields; and the Gujarat State Petroleum Corporation Ltd. (GSPCL) from the Hazira fields, which is being sold directly by them at market determined prices. Reliance Gas Transportation Infrastructure Ltd (RGTIL) has signed agreements for gas transportation with consumers and approached the Petroleum & Natural Gas Regulatory Board (PNGRB) for approval of the transportation tariff to be levied for ferrying Reliance's D6 block gas.

Most of the production of gas comes from the Mumbai high basin and the state of Gujarat. Large finds in the Krishna-Godavari Basin by ONGC and Reliance has bolstered the reserve estimates by 20 trillion cubic feet (tcf). The on-shore fields in Assam, Andhra Pradesh and Gujarat are other major sources of gas. Smaller quantities of gas are also produced in Tripura, Tamil Nadu and Rajasthan.

It is indeed a fact that the fertiliser and power sector uses more gas today than it did in the past. Presently, more than 10 per cent of our primary energy requirements are met through gas. However, India's domestic gas supply is unlikely to keep pace with demand, and the country will have to continue importing much of its gas needs, either via pipeline or as LNG. In order to give a thrust to the development of LNG trade in India, the Government has set up Petronet LNG Ltd., with some private participation. The capacity of the regasification plant in Dahej has been doubled and expansion work at Kochi and Dabhol terminals are underway. Shell had set up an LNG facility at Hazira. The LNG regasification capacity would reach a level of 20 million metric tonne per annum (mmtpa) in 2011-12.

India is leveraging its state power and diplomatic influence to advance Indian energy interests abroad. Oil diplomacy is being seen as a major tool for ensuring India's energy security along with acquisition of equity oil and gas overseas. Not only are most of these opportunities in risky zones (like Nigeria), there is also fierce competition amongst countries like China who are ready to secure these assets in astronomical amounts. There has been limited success so far, particularly in pursuing gas opportunities through overland pipelines.

In addition, unlike in the United States, the Government's efforts in augmenting gas supply by tapping unconventional resources, like coal bed methane, natural gas hydrates, and underground coal gasification, is still at its infancy.

In order to meet the unprecedented demand for gas, India has to not only continuously review its gas demand and supply options and improve its geological information and gas statistics, but it has to also formulate and implement policies and establish institutions that ensure transparency, fair play, and provide a level playing field. Though NELP has provided transparency in transactions, there has not been significant international interest. There is an urgent need for: (1) unified gas pricing; (2) a uniform gas utilisation policy; (3) development of a pipeline network and ensuring reasonable transportation tariffs; (4) expanding of LNG regasification capacity; (5) bettering port facilities (increasing drafts at major LNG hubs) and making the LNG shipping policy more attractive to enable foreign investment; (6) setting up of gas hubs; (7) increased use of gas for city gas distribution and transportation (replacing diesel with CNG) and (8) establishment of a transparent and equitable regulatory regime.

The Regulatory Regime

PNGRB was established in October 2007 under the provisions of the Petroleum & Natural Gas Regulatory Board Act, 2006 to regulate the refining, processing, storage, transportation, distribution and marketing and sale of petroleum, petroleum products and natural gas, excluding production of crude oil and natural gas. This step was taken consequent to the deregulation of the petroleum sector with effect from April 1, 2002. Some of the functions of the PNGRB is to *inter alia,* (i) register entities in order to: (a) market notified petroleum and petroleum products and natural gas, and (b) to establish and operate LNG terminals; (ii) authorise entities to lay, build, operate, or expand: (a) common carrier or contract carrier pipelines, and (b) city or local natural gas distribution network; (iii) regulate: (a) access to common carrier or contract carrier pipelines and city or local natural gas distribution networks, and (b) transportation rates for common carrier and contract carrier pipelines; (iv) monitor the price charged by a local distribution company and (v) lay down the technical standards and specifications in activities relating to the construction and operation of pipeline and infrastructure projects related to the downstream petroleum and natural gas sector.

The Centre-State Tussle

The Government of Gujarat had promulgated the Gujarat Gas (Regulation of Transmission, Supply and Distribution) Act, 2001 and raised objections to the inclusion of natural gas in the provisions of the above mentioned Act on the strength of Entry 25 of List II of the Seventh Schedule of the Constitution of India. The Union Government made a Presidential Reference to the Supreme Court of India in October 2001 for its opinion on the issue as to whether the subject 'natural gas' is a State or a Union subject. The Supreme Court, in its opinion dated March 25, 2004, held that Entry 25 of List II of the Seventh Schedule deals with 'gas and gas works' and the same relates to the manufacture of synthetic gas used for industrial, medical or other similar purposes. Further, it held that the definition of 'petroleum,' 'petroleum products,' and 'mineral oil resources,' contained in various national legislations, like the Oil Fields (Regulation & Development) Act, 1948 and the Mines Act, 1952 etc., and the national interest involved in the equitable distribution of natural gas among the states leads to the inescapable conclusion that 'natural gas' in raw and liquefied form is a petroleum product and hence, only the Union has the legislative competency to enact laws on the said subject.

Later, in 2005, the Gujarat Water and Gas Pipelines (Acquisition of Right of User in Land) Act, 2000 (GWGP) was challenged on the ground that natural gas, including LNG, is covered by Entry 53 of List I and hence, the State government has no powers to acquire land for laying pipelines for transportation of water and gas. The Gujarat High Court rejected this contention and held that GWGP is not in violation of the Judgment of the Hon. Supreme Court in Presidential Reference 1 of 2001. This issue has not got a quietus as yet.

Pricing of Gas

Gas prices are regulated and allocations to industry were determined by the Government. Some items like LPG continue to be heavily subsidised which were realised through higher than international pricing for petrol and diesel (lower than petrol).

Gas prices were last revised in September 1997, linking them to the prices of international fuel oils, with linkage progressively increasing to 75 per cent in 2000 (except the NE Region). From July 1, 2005, certain revisions were effected: (i) gas prices for APM gas produced by ONGC/OIL were revised on a provisional basis to Rs 3,200/mcm from Rs 2,850/mcm (fixed in October 1999) pending the Tariff Commission recommendations or till such time a regulatory mechanism is put in place—any additional or future gas produced by them will be sold at market-related prices in the context of NELP provisions; (ii) supply of non-JV/APM gas from ONGC/OIL to priority sectors of power and fertilisers (under the Gas Utilisation Policy) will continue against their existing allocations along with their specific end-users committed under court orders and small scale consumers having allocations up to 0.05 mmscmd at the revised rate; (iii) the gas price for transport sector (CNG) and small industries in Agra-Ferozabad and other small scale consumers having allocations up to 0.05 mmscmd would be progressively increased over the next 3 to 5 years to reflect the market price; (iv) gas produced by JVs will be sold at market determined prices, except 1 mmscmd of gas from Rava JV field in the Krishna-Godavari basin will continue to be supplied by GAIL under the existing arrangement; (v) Gas Pool Account will be limited to Rs 100 crore per annum against the present quantum of Rs 250 crore.

In 2010, the situation with regard to gas pricing continues to remain confusing, with approximately six pricing regimes being prevalent: (1) long-term LNG, (2) spot LNG, (3) gas sold under administered price mechanism (APM) to the North East, (4) APM gas sold to other regions, (5) Panna-Mukta-Tapti gas and (6) D6 (Reliance Industries Ltd) gas. The Government plans to move towards uniform gas pricing and may look at a pooled price mechanism which will ensure that similarly placed consumers do not have to pay different prices for the same gas without changing the producer price.

Gas Utilisation Policy

As per the production sharing contracts under NELP, gas from the NELP blocks should be marketed to consumers in accordance with the marketing priorities to be determined by GoI. Gas Utilisation Policy issued *vide* Press Note dated June 25, 2008 prescribes marketing priorities to supply gas from the KG-D6 field, the major source of gas in India. Under the said policy, city gas distribution (CGD) gets only the fourth priority, preceded by fertiliser plants, LPG and petrochemical plants and power plants. Refineries are pegged at the fifth slot, followed by other industries. Looking at the demand projections for various sectors (e.g. the demand projection for the power sector far outstrips the demand projection for fertilisers, year on year), the allocation priorities seem quite debatable.

The Mumbai High Court, in Reliance Industries Limited Vs. Reliance Natural Resources Limited (2009 (iii) Bom LR 2507) held that, '(i) the contractor has the freedom to sell production sharing contracts out of its share of profit gas," and (ii) the production sharing contract does not contain any provision "preventing the contractor from sharing gas at a price less than the price fixed by the EGOM out of the contractors entitlement of the cost of gas."' The matter is currently pending before the Supreme Court and the decision of the Supreme Court will have serious implications on the current policy on pricing and commercial utilisation of natural gas.

City Gas Distribution: Implementation & Distribution

The Government is targeting 200 cities to roll out the city gas distribution (CGD) projects—both compressed natural gas (CNG) and piped natural gas (PNG). A draft policy for development of natural gas pipelines, including city or local natural gas distribution networks, is on the anvil. However, regulatory and policy flaws appear to be hampering the growth of CGD in India. It took the Supreme Court, in its bid to control environmental pollution, to trigger robust usage of CNG in Delhi. The favourable economics of such conversions for end-users helped the switchover. Today about 40 cities and towns are covered by CNG. The Supreme Court, by its Common Order dated July 28, 1998 directed *inter alia,* (i) conversion of all city buses in Delhi to single fuel CNG mode by March 31, 2001; (ii) prohibition of plying of eight-year old buses, except on CNG or other clean fuels, by April 1, 2000 and (iii) GAIL to expedite and expand CNG supply outlets from 9 to 80 by March 31, 2000.

More recently, due to the non-notification of Section 16 of the PNGRB Act, there has been legal wrangling and judicial interference questioning the legality of the authorisations issued by PNGRB. PNGRB had issued CGD licences for six cities. Further, it had invited, by public notice dated February 13, 2009, another round of bids for seven cities—including Ghaziabad, in spite of the fact that Indraprastha Gas Limited (IGL) had obtained a 'no objection certificate (NOC)' from the Government of India for CGD for Ghaziabad, prior to coming into force of the PNGRB Act and when IGL's application for authorisation was pending before PNGRB. Later, PNGRB dismissed IGL's application by order/letter dated March 19, 2009. IGL challenged the said public notice and the said order before the High Court of Delhi. The High Court, by common order dated January 21, 2010 in W.P. (C) 9022/2009, held *inter alia* that, 'in view of non-notification of Section 16 of the PNGRB Act, PNGRB has no power to grant authorisation to entities which applied to it for laying, building, operating, or expanding city or local natural gas distribution networks.' This order has serious repercussions as a number of companies have already made significant investments in the CGD business on the basis of authorisation issued by PNGRB during the first round of the bids. The third issue is the question over dual regulators. The Government is reportedly setting up a National Gas Highway Development Authority (NGHDA) that would facilitate the development of long-distance pipelines for transporting gas. This may cause further complications to CGD unless the powers and functions of both regulators are clearly demarcated.

In conclusion, we can say that certain essential characteristics of the Indian oil and gas industry have not changed, despite piecemeal attempts at reform in the last decade or so. With an emphasis on inclusive growth, the twin features of excessive governmental control and divorcing pricing from underlying economic realities is unlikely to be abandoned in the near future. Despite some ad hoc structural changes and limited public offerings, the Government appears to be against privatisation of the Navratna companies and thus it is apparent that public sector monopolies are here to stay. The reform process was started with a view to providing an efficient, competitive, and market-friendly sector. Galloping demand, a declining production profile, volatile oil prices, and rapid development of newer technologies for deep water exploration compelled the Government to open up the sector to private participation. The NELP has indeed opened up opportunities for private players but it has failed to attract oil majors to invest in India. Most consider the sector over-regulated and non-transparent. Clearly, India needs to take hard decisions, not necessarily always determined by compulsions of political economy, but with a view to optimising the performance of the oil and gas sector. The reforms must be real and visible.

References

Business Standard (2010). Interview of Anish De, CEO, Mercados Energy Markets India Pvt Ltd: *Business Standard,* March 23.

FICCI-Mckinsey (2010). *Gas 2015: Unlocking Opportunities through Partnerships.*

Narayan, S. (2006). *Documenting Reforms: Case Studies from India.* Macmillian Publishers India.

9 Coal for Power

Issues and Challenges

R.K. Sachdev

Introduction

As on 31 March 2011 the installed electricity generation capacity was 1,73,626 MW, of which 93,918 MW or 54 per cent is coal based. Out of 78,700 MW capacity originally planned to be added during the Eleventh Five Year Plan period (2007-2012), 52,850 MW or 67 per cent is coal based. Therefore, it is very evident that our electricity generation programme will continue to be heavily coal dependent at least for the next three to four decades. During the Eleventh Plan period (first four years) over 32,500 MW has been added till March 2011 end.

In order to achieve economic growth of 8-9 per cent in terms of GDP, the Integrated Energy Policy Report has projected that in the year 2031-32, we have to have electricity generation capacity of 800,000 MW. Correspondingly, country's total coal demand has been projected to increase from ~ 730 million tonnes in 2010-11 to over 2,000 million tonnes in 2031-32. Even if we allow for the adverse impact of the economic slowdown of economy, the coal demand in 2031-32 would still be 1,800 and 2,000 million tonnes. Of this about 70 per cent of coal would go to power plants. This is a burgeoning task the domestic coal industry has to undertake, failing which the country's electricity generation programme would suffer and consequently the economic growth would also get adversely impacted.

Coal Availability

The power sector being the biggest single consumer, an adequate supply of coal has always remained a matter of concern both for power and coal industries as well as for the government. In the recent past many coal-fired power plants have reported critical stock positions. This gets further compounded by delayed monsoons, and leads to serious power shortages in many parts of the country. With a view to keeping a watch on the overall situation, the Cabinet Secretariat also periodically monitors the coal stocks at the power plants.

The Central Electricity Authority (CEA) monitors coal supply to all the coal-fired power plants in the country. Those plants which have less than four days of coal stocks are considered 'super critical', and those plants which have less than seven days stock are considered 'critical'. Reasons attributed to non-building of adequate stocks of coal *inter alia* included less supplies than approved quantities, increased demand by some plants upon achieving higher plant load factor (PLF), unloading constraints at some plants, less imports than targets and self-regulation by some plants to control inventory cost etc. Therefore, there is no doubt about the government's concern about inadequate availability of coal at the thermal power stations.

Coal Distribution Policy

The Ministry of Coal, in view of the directions from the Apex Court formulated this policy and it came into force on 18 October 2007. Salient features of the New Coal Distribution Policy (NCDP) as applicable to power sector are:

i. Erstwhile classification of consumers in 'core' and 'non-core' sectors has been dispensed with and each sector and consumers have been treated on merit, keeping in view the regulatory provisions applicable to them.
ii. Requirements of defence sector and railways to be met in full at the notified price.
iii. Power utilities, individual power producers (IPPs), captive power producers (CPPs) and fertiliser sector consumers would be considered for supply of 100 per cent of the normative requirement (further revised to 70 per cent) through Fuel Supply Agreements (FSAs) by coal companies at fixed price to be declared/fixed by them.
iv. Other consumers/actual users to be considered for supply of coal at 75 per cent of the quantity as per normative requirements through

FSAs by coal companies at fixed prices to be declared/fixed by them. Balance 25 per cent will have to be sourced from e-auction/import etc., as per their preference.

v. All existing linkage holders have to enter into an FSA with the coal companies.

vi. It has been proposed to replace the linkage system with a more transparent bilateral commercial arrangement of enforceable FSA.

vii. The Letter of Assurance (LOA) to be issued to new power sector consumers will have a validity of 24 months. The allottee of LOA is required to fulfill certain stipulated conditions and meet the milestones within this period and thereupon approach the coal companies for entering into an FSA.

viii. There is an emphasis on efficient use of coal and the existing norms of use are subject to change based on, in depth analysis by competent agencies.

The minutes of various meetings and discussions held among various government agencies like Ministry of Coal (MoC), Ministry of Power (MoP), Central Electricity Authority (CEA), Ministry of Railways (MoR), coal companies etc., that are available on the government websites are ample evidence of the efforts being made towards making the whole process of coal linkage, LOA and distribution of coal more transparent and objective.

In respect of power utilities and IPPs the over all positions of 221 LOAs is as follows (as on February 2011):

- All milestones achieved and coal supply commenced – 33
- Awaiting completion of milestones and valid through 2011/12 – 165.
- Applications pending due to incomplete documentation – 16.
- LOAs lapsed – 7.

Based on the inter ministerial decisions, power projects that are included in Eleventh Five Year Plan and are to be commissioned during the current Plan period or in early Twelfth Plan (2012–2017) period are being given priority for giving coal linkages. The pending cases seem to be over and above that of the power projects slated to be commissioned within the Eleventh Plan period.

Here it may be noted that Coal India Ltd. (CIL) has gone on record saying that despite their best efforts there would be a 'substantive' negative coal balance in the year 2011-12 as well as in 2016–17. They have further stated that in case more LOAs are considered for power plants, CIL should be allowed to decide annual contracted quantity (ACQ) under FSA in consultation with the CEA for each of the project, based on availability, generation target, logistics and other relevant factors. The coal availability has further been affected by various restrictions imposed by Ministry of Environment and Forests (MoEF) and their 'go' and 'no go' policy in respect of forestry and related statutory clearances.

In such a situation it is obvious that the concerned agencies are bound to scrutinise and apply stringent checks on each and every application seeking coal linkage/LOA under the NCDP. Such project proponents who are able to source their own coal either from captive blocks or from imports would definitely be better off.

Role of Standing Linkage Committees

In the Ministry of Coal, there are two Standing Linkage Committees for allocation and distribution of coal to the power plants.

- Standing Linkage Committee (SLC): Long-term (LT)
- Standing Linkage Committee (SLC): Short-term (ST)

Prior to introduction of the New Coal Distribution Policy (NCDP) in October 2007, SLC (LT) used to consider and approve requirement of coal at the planning stage. With NCDP in place, this committee now considers the requests for LOA from the consumers and recommends the same to the coal companies for further action in accordance with the provisions of NCDP.

With the implementation of NCDP 2007, coal dispatches to the consumers are now required to be ensured through FSAs. The institution of SLC (ST) for erstwhile core sector consumers namely power and cement plants, which was in force from 1979 onwards, has since been replaced by an 'FSA Monitoring Implementation Committee for Power Sector' with effect from June 2010. This Committee is to address the issues and disputes arising out of the FSAs between consumers (power utilities including IPPs) and coal companies covered within terms and conditions of the FSAs signed between the coal companies and consumers. (Ministry of Coal OM No. 2304/1/2009-CPD dated 17th June 2010).

E-auction

According to the government, the coal distribution through e-auction provides an access to coal for such consumers who were unable to source coal through the other institutional mechanisms for reasons like seasonality of coal requirement, limited requirement not warranting long-term linkage etc. It is also contemplated that e-auction could be a precursor to future spot coal markets in the country.

This logically leads to two conclusions: (1) a significant segment of coal consumers have no access to regular linkage/allotment channel to meet its coal requirement and (2) the notified price of coal is below the market benchmark. However, the e-auction route is not really meant for the power sector including the captive power plants that are serviced under the NCDP.

Projected Coal Demand

Coal demand as projected in the Report on the Integrated Energy Policy is given in Table 9.1.

Table 9.1

Projected Coal Demand

(Million tonnes)

Year / *Sector*	*2005-06*	*2006-07*	*2011-12*	*2016-17*	*2021-22*	*2026-27*	*2031-32*
Electricity (A)	310	341	539	836	1,040	1,340	1,659
Iron & Steel	43	43	69	104	112	120	150
Cement	20	25	32	50	95	125	140
Others	53	51	91	135	143	158	272
Non-elect. (B)	116	119	192	289	350	403	562
Total (A)+(B)	426	460	731	1,125	1,390	1,743	2,221

Note and Source:

- Figures for 2011-12 and 2016-17 are of the Working Group for Eleventh Five Year Plan's estimate and for 2031-32 are of the Integrated Energy Policy Report.
- Figures for intervening years have been extrapolated.

Domestic Coal Production Plan

Domestic coal production plan as envisaged by the Ministry of Coal (MoC) is given in Table 9.2.

Table 9.2

Domestic Coal Production Plan

(Million tonnes)

Entity	Tenth Plan	Eleventh Plan					Twelfth Plan
	2006-07	2007-08	2008-09	2009-10	2010-11	2011-12	2016-17
	Actual	Actual	Actual	Actual		#Projected	
CIL	361.02	379.49	403.73	431.00	431.32	447.00	664
SCCL	37.71	40.64	44.54	50.00	51.33	51.00	45
Captive	19.29	26.00	30.03	38.00	36.30	43.00	304 (?)
Tata Steel	7.04	7.21	8.95	7.20	7.02	7.00	7
Meghalaya	5.79	5.60	5.96	5.70	6.09	6.00	6
Total	430.85	456.40	493.21	531.90	532.06	554.00	1,026 (?)
*Demand	**450	**502	553	604	630	670 (?)	1,125 (?)
Gap	19.15	45.6	59.79	72.10	97.94	116	99 (?)

Note: * Demand given in the table is as originally projected for Eleventh Plan period.

** Assessed demand including imported coal.

Coal India has lowered their production target for FY 2011-12.

Coal Imports

Presently the country imports about 85 million tonnes of coal. Out of this, about 25 million tonnes is coking coal for the iron & steel industry. The balance is thermal coal used by power plants (50 per cent), cement industry (17 per cent) and other industries (33 per cent). Major sources of thermal coal imports are Indonesia, Australia and South Africa.

While PSUs are using Minerals and Metals Trading Corporation (MMTC) and State Trading Corporation (STC) as their procurement channels, the private companies are importing directly on their own through domestic and international trading firms. Recently, CIL has also been asked to import some five million tonnes to supplement the domestic availability.

Many Indian companies have acquired coal assets in Australia, Indonesia, Mongolia, Mozambique and South Africa etc. Quite a few including major PSUs like NTPC are seriously hunting for thermal as well as coking coal mines abroad. Coal Ventures International Ltd. (CVIL) a recently formed 'special purpose vehicle' promoted by five PSUs namely NTPC, NMDC, SAIL, RINL and CIL is also making concerted efforts to secure thermal and metallurgical coal mines abroad.

In view of the perennial shortages of domestic coal, there is a strong case for looking at a viable, transparent and competitive business model for import of thermal coal by public utilities, like NTPC, Generating Companies (GENCOs) and SEBs.

Since coal imports are set to increase this calls for more transparency and competition amongst the importing entities.

Present coal handling capacity at the ports is around 85-90 million tonnes per annum. This has to be augmented to at least 120 million tonnes per annum in next two years. Further is has to be doubled in the Twelfth Plan period.

Captive Mining Policy

Captive mining policy introduced in 1993 has also not provided much relief. For various reasons out of 208 blocks containing coal reserve of over 50 billion tonnes and with an aggregate ultimate production capacity of 560 million tonnes, only some 30 odd mines have commenced production. During 2010-11, total production outside the CIL and SCCL sources was only about 49.41 million tonnes. Captive mines contributed merely 36.30 million tonnes and the balance was from Tata's collieries and small mines in Meghalaya.

During Eleventh Plan period, the captive blocks were originally expected to contribute 104 million tonnes towards the total all-India coal production. However, at the present rate their contribution is not likely to be more than 43 million tonnes. This shortfall will further accentuate domestic coal availability in the country.

The overall position in respect of captive coal blocks is given in Table 9.3.

Table 9.3

Overall Position in Respect of Captive Coal Blocks

(As on February 2011)

Sector/End-Use	*Number of Blocks Allocated*	*Geological Reserve in Billion Tonnes*
Public Sector Companies		
I. Power	53	18.92
Power – UMMPP	12	4.85
II. Commercial mining	39	7.31
III. Iron & Steel	4	1.71
Sub Total I + II + III (A)	**108**	**32.79**
Private Companies		
Power	28	5.01
Iron & Steel	61	8.60
Cement	6	0.63
Coal-to-liquid	2	3.00
Others	2	0.39
Sub Total (B)	**100**	**17.63**
Grand Total (A+B)	**208**	**50.42**

The above stated figures reveal that over 18 per cent of total geological coal resource has been allocated for development of production capacity on captive mode.

More coal blocks are expected to be made available for allocation to captive end-users. However, since the number of applicants is increasing, and also with a view to making the allocation process more transparent, the Mines and Mineral (Development & Regulation) Act, 1957 has been amended for introduction of an auction system of allocation based on a competitive bidding process. Presently, government is reported to be in the process of consulting all the stakeholders for framing the rules for introduction of auction based allocation of coal blocks.

Issues pertaining to captive mining that require immediate resolution and attention of the government and other stakeholders *inter alia* include:

a. contentious issues to be resolved;

b. facilitating availability of geological data;

c. augmentation of exploration efforts;

d. expediting land acquisition and resolution of R&R issues;

e. forestry, environment and related clearances; and

f. getting mining leases etc.

In a significant number of cases lack of progress in the development of captive mines can also be attributed to the allocatees who are not making serious efforts to take things forward. The recent economic slowdown has also adversely impacted some of the downstream end-use projects because of a financial crunch.

Notwithstanding any of these reasons, if a country's coal demand has to be met, there is no option but to expedite the development of these blocks which were allotted hoping that the private developers would be in a better position to deal with various issues and bottlenecks and getting faster output of coal.

The government is reported to be working towards opening up of the coal sector without any restriction on the marketing of coal. This, however, would happen only after legislative approval, which is likely to take some more time.

Steps for Increasing Coal Production

In addition to the captive developers, the public sector coal companies that are contributing 83 per cent of total production also need immediate support in taking measures for increasing coal production from existing and new mines. Such measures *inter alia* include:

- Improving man, machine and capital productivity and a major thrust on underground mining needs immediate attention.
- It is well known that both government coal companies as well as the captive block allocatees are facing serious hurdles in getting mining leases and other statutory clearances.
- Role of the state governments in coal mining: concurrent development of coal production capacity, as in case of power plants.
- Production, quality and delivery system must be made strong and transparent.
- Facilitating commencement of mining in captive blocks of NTPC: lending personnel, contracting or JV.
- Coal imports for power plants.

Challenges Before the Indian Coal Industry

Institutional set up and management of the Indian coal industry has been a subject of debate every now and then. In recent past, the government had set up a committee to study the coal industry's problems and come out with recommendations with a view to making the coal sector more self-reliant in terms of technical management capability so as to function in a competitive environment. This committee has identified the following areas that need to be addressed:

- speeding up exploration by opening it to private sector;
- productivity improvement and increasing share of underground mines;
- institutional capability;
- coal washing and transportation;
- environmental approval issues;
- climate change issues;
- governance and regulation; and
- R&D in coal technologies and new resources.

India's Coal Resources

India's latest inventory of coal as updated by the Geological Survey of India is given in Table 9.4.

In recent past there has been a slant criticism of the reliability of India's coal resource base. In this context it must be mentioned that the Geological Survey of India (a 150-year old institution) gives the estimate based on their own data plus inputs available from various public and private agencies involved in carrying out coal exploration. The estimated figure of ~ 277 billion tonnes as on April 01, 2010 includes about 110 billion tonnes of proven resource. It is being said that if India's coal resources are re-estimated on United Nations Framework Classification (UNFC) methodology these would be much lower than the official stated figure. This is a wrong and non-tenable statement. A mere change in methodology of estimation will not materially change the total numbers. Only error margin in our estimation system, Indian Standard Procedure (ISP system) is to the extent that it does not account for the coal reserves that have been extracted so far. Even today if the total coal produced in the country in last 110 years is deducted from the total resource it will not make any material change in the total figure. Error would of < 1 to 2 per cent.

Table 9.4

India's Coal Resources

(As on 1.4.2010)

State	*Coal Resources in Million Tonnes*			
	Proved	*Indicated*	*Inferred*	*Total*
Andhra Pradesh	9,257	9,730	3,029	22,016
Arunachal Pradesh	31	40	19	90
Assam	349	33	3	385
Bihar	0	0	160	160
Chhattisgarh	12,441	30,230	4,011	46,682
Jharkhand	39,633	30,992	6,338	76,964
Madhya Pradesh	8,505	11,267	2,616	21,988
Maharashtra	5,360	2,984	1,965	10,308
Meghalaya	89	17	471	576
Nagaland	9	0	307	315
Orissa	21,507	32,074	12,726	66,307
Uttar Pradesh	866	196	0	1,062
West Bengal	11,753	13,030	5,071	229,853
Total	**109,798**	**130,654**	**36,359**	**276,810**

Source: Geological Survey of India. (Data may not add up due to rounding off).

Profile of Indian Coal Deposits

A quick look at the profile of Indian coal deposits would reveal the following:

a. Compared to other countries' coals, Indian coals contain high ash content, low volatile matter and have low sulphur.

b. Coal deposits are mainly distributed along the present day river valleys.

c. Deposits are of drift origin and high in inherent ash content.

d. There are 44 known coalfields located in the peninsular India containing 95 per cent of total resources.

e. Major coal deposits are confined to south and southeastern quadrant of the country.

f. Prognostic resource as on April 01, 2010 is 277 BT.

g. About 110 BT is of proven category.

Regulatory Aspects

The present institutional structure is a near monopoly of the government. Regulatory position can be described as below:

- Near monopoly industry structure.
- Eighty-three per cent coal production comes from government-controlled mines.
- Central government owns, regulates and operates.
- Primary regulatory authority is the Ministry of Coal.
- Safety aspects are regulated by Directorate General of Mines Safety (DGMS).

While on one hand the requirement of an independent regulator has been under active consideration of the government, there are also some divergent opinions on the need of having an independent regulator.

Those who do not favour (a very small number!) the institution of an independent regulator are skeptical about CIL as a Maharatna company might overshadow the very objectivity of the purpose for which the regulatory institution is created. They are of the opinion that the government should in the first instance strengthen the coal market with more private players so that the production contribution from privately-owned coal mines should become significant say at least 100 million tonne. At that stage they feel that the regulator would able to function independently and serve the purpose more effectively.

Those who are in favour, support the need of an independent regulator because:

- Government controlled producers do not enjoy confidence of consumers and private investors.
- Present overseeing by the Ministry of Coal (MoC) is too general to operate with required degree of expertise.
- Level playing field is not available to private developers.
- Transparency in price fixation is not there.

- Captive block owners are always at the receiving end.
- Captive developers are not allowed to share common infrastructure like railway sidings with the government-owned companies even after making payment for the same.

The Government has got the whole issue of need or otherwise of an independent coal regulator examined by the Administrative Staff College of India (ASCI). Based on the ASCI report, a draft Bill is under consideration of the Government. Based on the overall consensus, an independent regulator for the coal sector has already been decided. Modalities of appointment are reported to be under way.

Domestic Price Fixation

The Government of India deregulated the prices of non-coking coal of grades A, B and C, coking coal and semi/weakly coking coal on 22.03.1996. Subsequently, on 12.02.1997, the Government of India deregulated the prices of non-coking coal of grade D, hard coke and soft coke and also allowed CIL to fix coal prices for grades E, F and G till January 2000 once every six months by updating cost indices as per escalation formula contained in the 1987 report of the Bureau of Industrial Costs & Prices. With effect from 01.01.2000, CIL was free to fix the prices of such grades of coal in relation to the market prices.

Accordingly, CIL and Singareni Collieries Company Ltd. (SCCL) revised the coal prices keeping in view the impact of rise in input costs, wage cost and other such factors including the demand of any particular variety of coal like metallurgical coal for which border price indicators are available. Last price revision was made effective by CIL on 27.02.2011. There is no subsidy component in coal pricing on an overall basis.

Coal Quality

Quality is an integral part of coal supply chain management. The scenario that prevailed till the recent past and is still continuing in some cases has been as follows:

- Power plants complain of high ash content, inconsistent quality and size.

- Over 87 per cent of coal is produced from mechanised open pit mines.
- There is a pronounced degree of contamination during mining.
- Multiplicity of supply sources in many cases further adds to the problem of inconsistency in the quality of coal.
- A number of committees set up by the government (between 1978 and 2003) have recommended washing of thermal coal.
- From June 2001, government had mandated the use of coal of ash content below 34 per cent for power plants located over 1,000 km from the coal supply source and those located in urban and environmentally sensitive areas.
- All the power plants that are covered by the mandate are not getting coal of desired quality.

After a long and protracted debate that lasted for over 30 years 'whether to wash' or 'not to wash' thermal coal for power plants, the government has decided that all coal supplied to power plants (except for pit-head stations) should be washed at the mine mouth. Accordingly, CIL has announced the following action plan:

- CIL will supply washed coal to all thermal power plants, except the pithead ones.
- CIL plans to set up 20 coal washeries of aggregate capacity of 111 million tonnes in next 2–3 years and establish capacity for 100 per cent coverage by 2017.
- Capacity establishment will cost CIL~ Rs 4500 crore or USD 1.0 billion.
- Washeries would be set up on 'build, operate and maintain' (BOM) basis through global competitive bidding process using state of the art technology.
- CIL will provide funds, land, water, power and assured coal supply.
- CIL will market the coal.

Global tenders for six washery plants with aggregate capacity of 40 million tonnes per annum have so far been issued and two plants have been awarded. The remaining four are reported to be in advanced stage of evaluation for construction work. In addition, CIL, based on request for qualifications (RFQs) has already completed the short-listing of parties for the remaining washeries.

As a precursor to the recent initial public offering (IPO) that helped CIL to garner over Rs 1,5200 crore from its maiden public issue, the company having realised the commercial importance of 'coal quality' had in its pre-IPO road-shows declared that it would set up adequate washing facilities to ensure the quality of coal supplies to all consumers.

Going by the foregoing developments one can hope that the contentious issue of coal quality will be a thing of past in coming five to seven years.

Infrastructure Related Issues

Hitherto, the development of new coal mines was taking place wherever transport infrastructure for evacuation of coal and its further transportation to various designated destinations could be managed without much of a problem. With the increased demand, more and more new and far-flung coalfields are being taken up for development to meet the increasing demand of coal in the country. Initially such developments can go along with road transport. But road haulage is not easy due to lack of road infrastructure of adequate strength. This highlights the need for development of Railway facilities for all such locations.

Similarly, for handling and transportation of increasing volumes of imported coal, integrated port and railway infrastructure has to be established.

Impact of Forest Regulations

Ongoing debate on 'go' and 'no go' categorisation of coal bearing areas is already having serious negative impact even on the current domestic production of coal. As a result CIL has already scaled down their target for next FY 2011-12 to 447 million tonnes from the original figure of 520 million tonnes. Similarly is the situation with the captive block owners. Domestic coal industry has suffered a serious dent it terms of its coal development plan goals. This has directly impacted the power sector. Most of the coal based power stations are suffering from coal shortages.

Coal mining being a location-specific activity, government must adopt a pragmatic approach based on international experience. After all we are not the only country carrying out coal mining in forest areas. Government as well the industry have their roles to play and if a judicious approach is adopted, there is no reason why we can not overcome the ongoing syndrome which has brought the development of the coal industry to a virtual halt.

Coal and Climate Change

The adverse impact of increased CO_2 emissions from increasing use of coal in thermal power stations and other industries is well known. Climate change issues are being debated globally particularly in the context of increasing growth rate in countries like India and China where energy supply is heavily coal-dependent.

In this backdrop, it is highly prudent that in the new power plants, India opts for clean coal technologies, which are more efficient and less polluting in terms of CO_2 emissions.

Adoption of 'super critical' steam parameters in Ultra Mega Power Projects (UMPPS) and other plants is definitely a step forward. It is time to decide not to allow any new coal based power plant with sub-critical parameters.

Old and inefficient power plants should both be renovated and modernised to improve their heat rate and wherever such improvements are not considered feasible, they should be phased out.

In this context, it is important to mention that the CEA and the Bureau of Energy Efficiency (BEE) along with GTZ (Germany) and UNDP have brought out various benchmark norms for CO_2 emissions. Power industry must adopt more efficient and low carbon generation technologies.

Conclusions

1. With its low-cost and wide spread resource base, coal will continue to contribute to economic growth and social development across the globe.
2. Global coal based generation will grow from 6,000 TWh (10^9 Kwh) in 2,000 to 12,000 TWh in the year 2030.

3. Coal accounts for nearly 58 per cent of India's commercial energy consumption and over 70 per cent of coal goes for generation of electricity.
4. Widening demand-supply gap is an immediate concern and steps should be taken for increasing domestic coal production for long-term energy security.
5. Accelerated development of captive coal blocks is must. For this, outstanding issues must be resolved early.
6. Government must do more to remove the roadblocks and treat allocatees of captive blocks as 'partners in progress'.
7. Captive block allocatees have direct responsibility to ensure that they meet the agreed timetable for commencement of production.
8. A strong domestic coal production and delivery system would be imperative if the country has to achieve the goal of energy self-sufficiency and long-term energy security.
9. Coal imports are set to increase. This calls for securing coal prospects abroad and development of port capacity with matching inland transport infrastructure.
10. An independent coal regulator is required to create confidence in the mind of private investors and provide them a level playing field.
11. To contain carbon emissions, the coal-energy chain has to be clean and environmentally acceptable.
12. At the same time, the impact of the increased use of fossil energy sources has to be enjoined in the policy framework so as to guard against any negative environmental trade-off in future particularly in the context of global climate change concerns.
13. Some of the important issues which the government and the coal industry need to address on post-haste basis are:
    a. We should make serious efforts in enhancing our extraction percentage from the operating and new mines.
    b. We should not make small blocks as this leads to 'sterilisation' of coal resource in the 'barriers' in-between adjoining mines.
    c. We should adopt more pragmatic approach in allocating coal blocks. Excessive sub-blocking merely to satisfy 'hunger' of large

number of individual companies to operate coal mines is leading to that.

d. We should allow, rather force that captive block owners to commence production immediately irrespective of the preparedness to consume in their own captive end-using unit for any reason. Coal thus produced should be allowed to be sold/ supplied to the designated 'end-users' who are facing shortages like many power stations.

e. If this allowed, we will have at least 50-60 million tonnes of additional coal available within next two years, subject to other clearances etc.

f. Too much reliance on imports and/or overseas coal acquisitions could prove a serious barrier in the development of domestic coal industry.

g. Our 'go' and 'no go' policy barrier is only a mental block. We need to think out of the box. Easily and cost-effective solutions can be found based on experience in Appalachian coal mines in the US and Australian coal mines.

h. To achieve this, the government must be open to new ideas and the industry serious and sincere in implementing the required environmental restoration measures.

References

Coal Controller's *Provisional Coal Statistics, 2010-11*.

Websites of the Ministries of Coal, Power and the Central Electricity Authority, Coal India Ltd. etc.

Author's own published and unpublished papers/reports and other similar published sources.

Section V

Open Access Scenario

10 Open Access

Powering the Economy

Raghvendra 'Raghav' Upadhya

India's growth story can be hamstrung by the lack of basic infrastructure support in the energy, power, transport, water, food and health sectors. The creation of infrastructure, through its direct and indirect effects, has a significant impact on poverty reduction. India's economic development will be boosted as it has been from the time the NDA government invested in the Prime Minister's Golden Quadrilateral Roads Programme as spending on infrastructure has large multiplier effects.

The Indian Power Sector

The Need for Power Generation by the Private Sector

Power being a case in point, an immense shortfall in the supply of power as compared to demand, massive losses in transmission & distribution (T&D) to the extent of 25.47 per cent, lack of adequate metering systems, rampant corruption and power theft, have characterised the state of affairs of the power sector in India.

Table 10.1

Power Supply Position 2010-11 (February 2011)

Region	*Energy (MU) Requirement*	*Deficit %*	*Peak (MW) Demand*	*Deficit %*
Northern	19,181	-6.3	33,717	-9.7
Western	22,979	-13.3	40,502	-16.8
Southern	19,521	-4	31,680	-4.3
Eastern	7,054	-4.5	12,972	-5.9
North Eastern	766	-5.5	1,665	-6.8
All India	69,501	-7.8	120,536	-10.2

Source: *www.cea.nic.in*

Power generation in India was largely driven by the government sector, which till date, accounts for almost 85 per cent of the power produced in the country. However, state government authorities had weakened themselves financially over the years by not controlling high T&D losses and rampant theft of electricity. They had also allowed short-term political compulsions such as highly subsidised power to interfere with long-term economic considerations. Thus, they had very limited funds for setting up new power projects. They were also unable to receive adequate funding to set up such projects owing to their poor track record in financial management. The inability of state authorities to add generation capacity due to these reasons was sought to be remedied by Central government authorities in the 1960s-70s and onwards leading to NTPC and NHPC being set up to bolster power generation. However, by the 1990s it was clear that private and foreign participation were also required to fund and develop new generation capacities in order to match the rapidly increasing demand.

Creation of Electricity Markets

New power generation to the extent of 77,000 MW needs to be added by 2012 (end of the Eleventh Plan), which requires an approximate cost of upwards of USD 300 billion. However, resources of power generation are unevenly dispersed across the country because of the concentration of coal resources in the Bihar, Jharkhand, Chhattisgarh, Madhya Pradesh, Orissa belt. Further, electricity is a commodity that cannot be stored in the grid where demand and supply have to be continuously balanced. Thus, given the amount of investment required coupled with the risks associated with the sector, the Electricity Act, 2003 sought to stimulate investments through creating an electricity market so as to de-risk the single buyer model. In a large country like India, shortages in parts of the country at some time might witness surpluses elsewhere and trading can help the surpluses to be moved to reduce some of the shortages. The monopsonistic buyers in the Indian context: the electricity boards which are controlled by state authorities were incurring huge losses and had no money to pay for their purchases of electricity. The electricity boards were therefore not acceptable as buyers unless there were guarantees of payment, to foreign and private lenders for development of additional capacity projects. There is and has been a need to have an alternative default option for the sale of power from a paying buyer to de-risk these projects.

The creation of a wholesale electricity market would also help enable the widely distributed and rapidly increasing demand requirements of the country to be met in an optimum manner by customers willing to pay. The National Electricity Policy, 2005 recognised the need for encouragement of a trading market and realised that it would help finance capacity additions. Even the National Tariff Policy 2006 recognised that it is essential to attract adequate investments in the power sector by providing appropriate return on investment as budgetary resources of the Central and state governments are incapable of providing the requisite funds. It recognised that it is equally necessary to ensure availability of electricity to different categories of consumers at reasonable rates for achieving the objectives of rapid economic development of the country and improvement in the living standards of the people. McKinsey & Company's 'Powering India: The Road to 2017' Report on the future of the sector asserts that trading activity requires to be encouraged, which would stimulate investments, especially in peaking power plants, which are urgently needed for meeting the ever rising peaking demand shortage in this country (nearly 40 per cent of India's demand today is non-baseload) and that there is a need to create a deep and well functioning wholesale electricity market. Considering that power trading has been in existence for over seven years, even now the volume of trading has remained around 10 per cent of the total power supply.

Box 10.1

Open Access in the Indian Power Sector

The Electricity Act was passed in 2003 by the Parliament of India to bring in transparency, efficiency, accountability, and to ultimately create an electricity market and provide consumer choice in the power sector. A key element of the Act was to incentivise power production in which the provision of non-discriminatory, open access to the grid for power producers was to play a major role.

Definition of Open Access in the Electricity Act 2003

Open Access means the non-discriminatory provision for the use of transmission lines or distribution system or associated facilities with

contd...

...contd...

such lines or system by any licensee or consumer or a person engaged in generation in accordance with the regulations specified by the Appropriate Commission-Part I, Definition (47) of the Act.

Recent Experiences in the Denial of Open Access

1. Citing 'public interest' the State Government of Karnataka passed an order on 30 December 2008 requiring all generators to supply all exportable electricity generated to the State Grid, in view of the 'extra-ordinary circumstances'.
2. Tamil Nadu passed an order to the same effect on 27 February 2009 citing 'public interest'.
3. Orissa passed a similar order to the same effect on 22 April 2009 citing 'public interest' and 'extra-ordinary circumstances'.

Enforcement of the Grid Code and Open Access

1. The threat of a grid collapse being very real, in May 2009, the CERC fined Tamil Nadu, Karnataka and Andhra Pradesh, among other states on account of Indian Electricity Grid Code Violations.
2. On 23 June 2009, the Minister of Power, Government of India, Sushil Kumar Shinde urged state ministers to follow the Open Access norms in letter and spirit as power capacity addition in the private sector are dependent on it.
3. On 7 July 2009, Sushil Kumar Shinde announced the intention of the Ministry of Power, Government of India, to amend Section 11 of the Electricity Act, 2003 to allow IPPs to sell power to consumers of their choice, doing away with the current intervention of states.
4. On 18 May 2011, the CERC wrote to the Secretary, Ministry of Power, Government of India, noting that it has issued orders ruling that open access has to be granted in accordance with the regulations under the Electricity Act and the same cannot be interfered by (state controlled) SLDCs on the basis of orders under Section 11. It also noted that there was an urgent need to settle the correct legal position of such directions by the State under Section 11 or 108 or any other provisions of the Act.

...contd...

...contd...

Status of Applications Received for Open Access in Distribution (31-3-2009)

States	*Received*		*Approved*		*Implemented*	
	No.	*MW*	*No.*	*MW*	*No.*	*MW*
Andhra Pradesh	9	130.5	2	44	2	44
Chhattisgarh	14	332.8	6	66	5	53
Gujarat	15	871.4	15	871.4	15	871.4
Madhya Pradesh	29	59.6	29	59.6	29	59.6
Maharashtra	46	13170.5	41	13122.5	5	88
Rajasthan	30	271.1	12	165.3	12	165.3
Tamil Nadu	12	1764	0	0	0	0
Other States (*)	19	874.2	11	133.9	8	113
Total	174	17474.1	116	14462.7	76	1394.3

Note: (*): Other states where open access is permitted include Haryana, Himachal Pradesh, Jharkhand, Kerala, Orissa, Punjab, Uttar Pradesh and West Bengal.

Source: Forum of Regulators.

Impact of the Denial of Open Access

Transmission Perspective

Power transmission systems in the country are largely a natural monopoly. Also, it is a very capital intensive proposition to lay such lines across large distances, connecting power surplus entities to power deficit ones. The power transmission sector thus has significantly high entry barriers and little or no competition. In order to prevent the misuse of monopoly power, where the owner of the transmission or distribution system can negotiate unfairly low prices of electricity purchased from power producers or unfairly high prices of power sold to the consumer, there is a need to provide non-discriminatory, open access to such facilities to private power generators, sellers and buyers, enabling them to transact power with other buyers and sellers of their choice.

What are the social and economic costs of a monopoly such as a state government controlled entity denying open access in the power transmission sector? Economic theory and practice suggests that it is not just the higher profits that a monopoly can make on account of higher prices for its services, which is the reason why monopolies are generally not desirable.

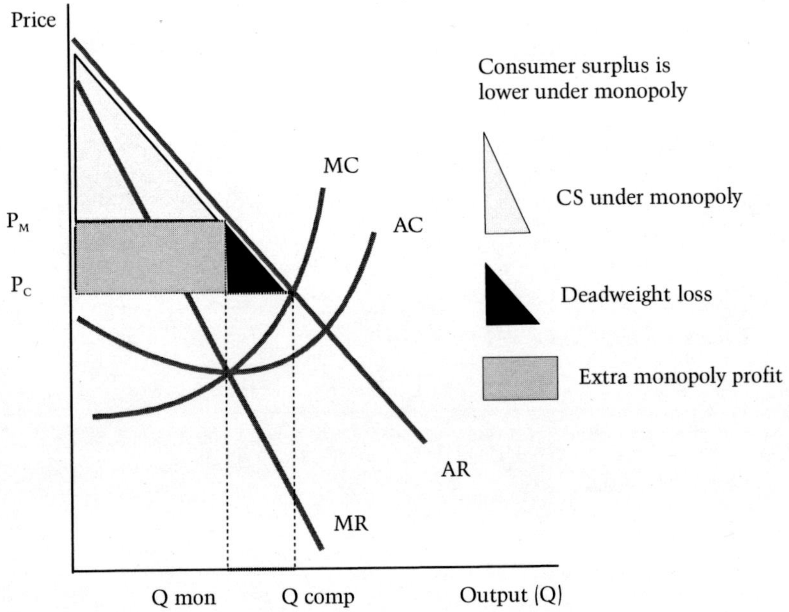

P_M – Price under monopoly;
P_C – Price under perfect competition;
Q_{Mon} – Quantity of output under monopoly;
Q_{Comp} – Quantity of output under perfect competition;
MC – Marginal cost;
AC – Average cost;
MR – Marginal revenue; and
AR – Average revenue.

Image Source: http://tutor2u.net/economics/content/topics/competition/efficiency.htm

It is the deadweight loss to society that is a concern which occurs since lower quantities of output are produced, transacted and hence consumed under a monopoly than the alternative: perfectly competitive markets. This implies that the denial of non-discriminatory, open access by state utilities to private power developers ensures lower levels of power are transacted (generated/transmitted/traded/sold to non-state buyers), implying a lose-lose proposition for private power developers and the state transmission utilities themselves.

In case there is no spare capacity available on transmission lines and alternative routes are not available, open access may be denied by the transmission utility. In the long term however, market forces should ensure other entities are incentivised to set up transmission lines on crowded, remunerative routes, thereby providing competitive alternatives to the monopoly. However, such private investments in power transmission, which are much required, have not fructified for various reasons.

Private Sector Perspective

Power plants are long gestation, high risk projects. A committed power off-take such as a power purchase agreement (PPA) and long-term fuel linkages are just the bare necessary requirements to provide surety of return on investments. If the setting up of a new power plant is to become a project attractive enough for investors and developers, both domestic and international, to invest money, technology and resources; what is really required is the freedom to sell to buyers who can pay more remunerative prices than monopsonistic state electricity boards.

About 95 per cent of power transactions in India are of power owned and controlled by government entities and sold at a regulated price under long-term PPAs. In order to subsidise the supply of power for free or at a nominal rate to farmers, politically significant groups of consumers and households, this regulated price is kept low. While such low power procurement prices can be used for old and nearly fully depreciated power plants set up decades ago and/or funded by the government, they are economically unviable for new power projects that in addition to the lack of open access may face fuel supply and cost uncertainties and high cost financing. An exception to this may be the Ultra Mega Power Project (UMPP) segment where projects are awarded to the lowest bidder as a well packaged deal with land acquisition, environment approvals, financing, fuel linkages, fiscal benefits and PPAs all worked out as part of a comprehensive public-private partnership (PPP) package deal. Other power projects are not as well endowed.

Consumers' Perspective

The Electricity Act, 2003 also provided consumers with a choice: they could choose between various suppliers of electricity. Denial of open access deprives consumers of this choice. It also compels them to rely on costlier power from alternative means such as liquid fuel based power. Since there is no incentive for the development of new power projects, even as demand for electricity increases, there is a yawning supply-demand mismatch, resulting in growing power outages of 6-12 hours a day in many areas. This constricts industrial growth which in turn negatively impacts related sectors. Till date only a handful of transactions have been granted under Section 42 (2).

Consumer and Private Producer Perspectives and Interconnections

The Electricity Act, 2003 requires a surcharge for open access, to meet some of the costs of cross-subsidising special interests like free power to farmers. The regulators of some states determined this surcharge at such high levels that transacting electricity using open access was unaffordable.

The Figure 10.1(a) shows the open access charges of select states. Figure 10.1(b) shows the tariff that electricity distribution companies (discoms) charge industrial consumers.

High charges make the use of open access unviable. High tariffs charged from industrial consumers similarly restrict expansion of industrial output. On the whole, new generation capacities are dis-incentivised from being set up, captive power units with surplus power similarly have no incentive to sell, power surplus areas stay stranded from power deficit areas, lesser quantities of electricity are transacted than the market clearing equilibrium, long term prices of electricity and power outages keep rising.

The Electricity Act, 2003, specifically states that governments can subsidise consumers as long as they make a provision in their budget and pay the subsidy up front. However, this is not happening for various reasons, political and otherwise, leading to imperceptible short term gains and mounting long term deadweight losses.

Figure 10.1

*(a) Open Access Charges (Rs/kWh)**

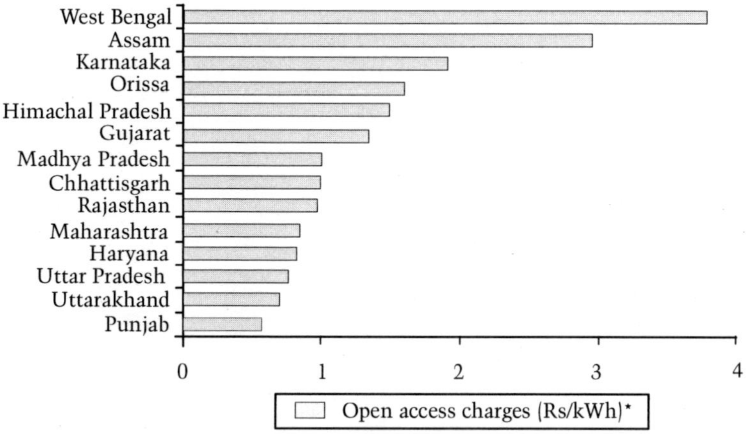

*(b) Tariff (Discom)***

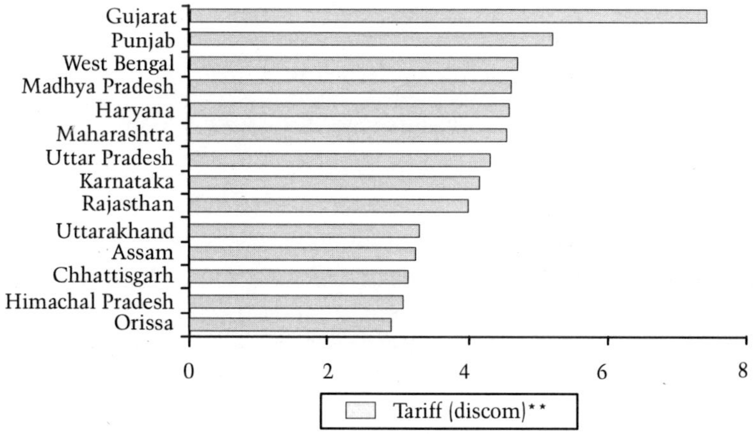

Note: * OA charges for a consumer of 5 MW at 11 KV (33KV in some (cases) seeking OA for a month. This includes transmission & wheeling losses calculated assuming power purchase cost as Rs 4/ kWh.

** Tariff for an embedded consumer of 5 MW at 11 KV (33KV in some cases).

Source: India Electricity Market Outlook 2008 by KPMG.

Granting of Non-Discriminatory Open Access: Incentivising Generation

The importance of power pricing influencing the internal rate of return (IRR) of a power project is shown in Figure 10.2.

Figure 10.2

Internal Rate of Return (IRR) of a Power Project

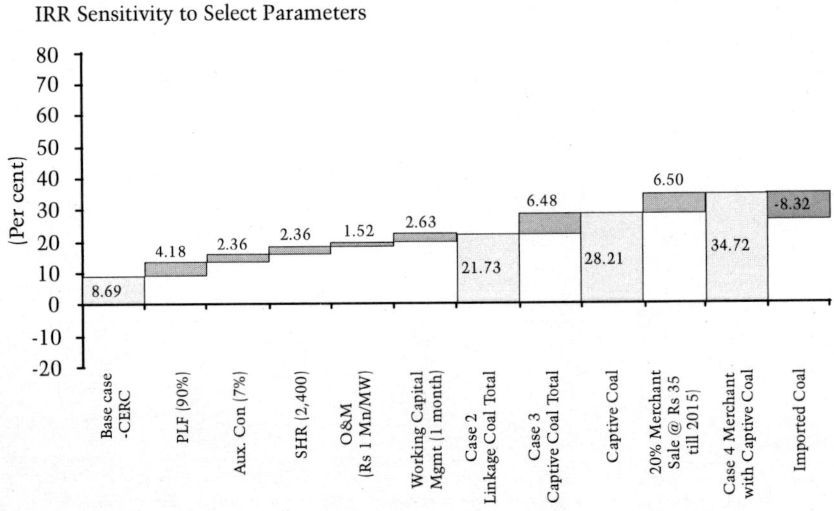

Source: KPMG.

The 'base case' of Central Electricity Regulatory Commission (CERC) mandated cost plus return figure provides an IRR of 8.69 per cent. Assumed values of variables such as plant load factor, auxiliary consumption, station heat rate, expenditure in operations and maintenance etc., are given as above. The IRR increases each time a variable is added. An assumption of even 20 per cent merchant sale from total power generated, provides a healthy increase of 6.50 per cent in IRR. This can greatly help in offsetting the impact of other business variables such as fuel uncertainties in the current economic condition (imported coal can reduce IRR by 8.32 per cent). Captive power can also be sold to offset the impact of an economic recession, where the sale of industrial output of the parent organisation has slackened.

The recent initiative of the Ministry of Power, Government of India, proposing the amendment of Section 11 of the Electricity Act, 2003, may be of help in implementing 'open access' as already mandated in the act, in letter and spirit. However, other sections of the Act can also be misused to deny open access. These are required to be amended as well. Recommendations for the same are as below:

Recommendations for Reforms in the Electricity Act, 2003

Invocation of Section 11 has led to instances wherein the state governments have misused the power granted under Section 11 resulting in the denial of open access. The present section is an attempt to highlight the possible changes that could be brought about within the current legal framework so as to ensure that the provisions of the Act are not misused.

1. Amendment to Section 11: 'Appropriate Government'

It is recommended that Section 11 be suitably amended to ensure that the state governments do not arbitrarily invoke the said section to suit the provision according to their whims and wishes. The word 'Appropriate Government' may be amended to 'Central Government'. Such an amendment shall provide an authority on the Central Government to issue directions to the generating companies in case of any extraordinary circumstances arising in the states and may also decide upon the issues to offset adverse financial impact in pursuance of such directions.

2. Appropriate amendment of the Act Resulting in Restricting the Power of the State Governments to Misuse Sections 11 and 108

It is stated that Section 11 and section 108 of the Act can be suitably amended to ensure that the state governments do not arbitrarily invoke the said sections leading to the denial of open access. Some of the changes which could be brought within the Act to ensure that the state governments do not misuse their power under the Act are as follows:

a. Amendment to Section 11 and 108

It is stated that both sections 11 and 108 could be amended so as to enhance the role played by the state commissions before a state government makes any direction under the said sections. Such an amendment could be brought in line with Section 12 of the Delhi Electricity Reforms Act, 2000 (hereinafter 'DERA'). Section 12 of the DERA provides as follows:

Section 12- General Powers of the Government

(1) In the discharge of its functions, the Commission shall be guided by such directions in matters of policy involving public interest as the Government may issue from time to time.

(2) If any question arises as to whether any such direction relates to a matter of policy involving public interest, the decision of the Government thereon shall be final.

(3) The Government shall be entitled to issue policy directions concerning any subsidy to be allowed for supply of electricity or any other infrastructure services to any class or classes of persons;

Provided that the Government shall contribute an amount to compensate the Board or any company affected to the extent of the subsidy granted. The Commission shall determine such amounts, the terms and conditions on which and the time within which such amounts are to be paid by the Government.

(4) The Government shall consult the Commission in relation to any proposed legislation or rules concerning any policy direction and may take into account the recommendations made by the Commission.

A bare perusal of sub-section (4) of Section 12 of DERA indicates that the Government shall consult the Commission in relation to any proposed policy direction. Such a consultation would ensure that the concerns of the Commission (which in effect is mandated to safeguard the interests of the consumers of electricity) are brought on record before the Government makes any policy decision. This would also ensure that before the Government makes any decision, it has the advice of a body which has the necessary technical competence.

Therefore, it is stated that both sections 11 and 108 could be amended to add a sub-section which ensures that before the Government makes any policy or issues any direction, it consults the Commission and has with it the necessary feasibility study before it arrives at any conclusion.

Consultation would ensure that compensation issues are taken into account at the time of issuance of the notification itself, as the Commission has to consider compensation issues, it may be able to spell out compensation principles at this stage itself. This ensures certainty to the Generating companies (GENCOs) that compensation claimed will not be brushed aside when the same are preferred under Section 11(2) of the Act.

From another perspective, the consultation may be used to specify, if possible, the basis of compensation in the notification itself. The consultation would ensure that the principles become actionable as mentioned above.

b. Amendment to Section 110

It is stated that Section 110 of the Act which establishes an appellate forum namely the Appellate Tribunal for Electricity (hereinafter 'ATE') for hearing appeals filed against the orders of adjudicating officer or the Appropriate Commission under the Act could be suitably modified to ensure that the directions issued by the Appropriate Government can also be impugned before the ATE. Such amendment could be done by providing the ATE with the jurisdiction to entertain an appeal against a direction of an Appropriate Government.

It is stated that the need to regulate the Government is justifiable as it fulfills the object of the Act, namely, regulation of the sector on economic, transparent and independent principles as set out under the Act. The accountability of stakeholders in the sector through a review of their actions by competent regulators is one of the touchstones of the Act and would be facilitated through the proposed measure. Any intervention by the Government other than through a regulated manner can only be as provided for under the Act, such as under Section 65 of the Act, which deals with the provision of subsidy.

The Parliament through the Act made a conscious decision of freeing the tariff related issues and regulation of the sector from Government control and vesting the same in competent and independent bodies such as the regulatory commissions to enable the efficient growth of the sector. Therefore, the objective of the Act ought not to be defeated through the use of the aforementioned Section in a manner which runs contrary to the objective of the Act.

Such an amendment would ensure that a forum like the ATE which has the necessary competence to adjudicate matters concerning the electricity sector is provided with the jurisdiction to judge the reasonableness of the orders of a state government. The ATE as a forum has the perfect blend of legal and technical members which would ensure that the orders of the state government are reviewed not only from a mere technical standpoint but also from a legal perspective.

The ATE is also not as structured with backlog as courts in India and the disposal mandate of the Act of 180 days would ensure timely adjudication of the challenge which is essential to tariff fixation and the financial health of the electricity sector.

Since the orders issued by the state government have serious and immediate consequences, the proposed amendment would also guarantee a speedy disposal of cases leading to an overall development of the electricity sector as a whole.

3. Appropriate Interpretation of the Provisions of the Act

It is stated that if the sections of the Act are interpreted in line with the spirit of the Act, the current debate would never have arisen. It is because of the fact that the provisions are interpreted in an arbitrary manner by the state governments that the current troubles arose.

It is stated that keeping in mind the general scheme of the Act, the Appropriate Government's power to issue directions should be only persuasive in nature and in no way should a direction issued by the Appropriate Government usurp the powers of the Commission. This interpretation also finds support in Delhi High Court's judgment in Gajendra Haldia *v.* State of NCT of Delhi and Ors. [2007 ELR (DEL) 1429] wherein the court with respect to Section 12 of the DERA observed as follows:

> However this Court directs that the GNCTD will hereafter, in exercising its powers under sections 12(1) and 12(2) DERA, abide by the mandate of the DERA and not reduce the role of the DERC to a mere consulted whose views do not matter. Clearly, the GNCTD cannot under the pretext of exercising its powers under section 12 DERA take over the function of determining the tariff as it has in the first phase.

From the above observation of the Hon. Delhi High Court, it is amply clear that the views of the concerned Commission should be considered and that the state governments cannot issue directions which would amount to the usurpation of powers of the Commission.

Therefore, it is stated that if the Act is amended as mentioned in the preceding paras, the Appropriate Commissions could have a more pro-active role to play in policymaking which would ensure that the interests of the consumers are safeguarded and the Appropriate Governments do not have unfettered discretion in instances of policymaking.

4. Regulations in Pursuance of Section 11(2)

Section 11 of the Act empowers the Appropriate Government, in extraordinary circumstances to issue directions to a generating company to operate and maintain any generating station. The relevant part of Section 11(1) reads as follows:

> The Appropriate Government may specify that a generating company shall, in extraordinary circumstances operate and maintain any generating station in accordance with the direction of that Government.

From a bare perusal of the above provision, it is amply clear that the Appropriate Government may in extraordinary circumstances issue directions to a Generating Company to operate and maintain a generating station.

It is stated that these directions can have possible adverse financial impacts. Sub-section (2) of Section 11 empowers the Appropriate Commission to offset such adverse financial impact. Section 11(2) reads as follows:

> The Appropriate Commission may offset the adverse financial impact of the directions referred to in sub-section (1) on any generating company in such manner as it considers appropriate.

From a plain read of the above provision it is amply clear that the Appropriate Commission is empowered to offset any financial impact that may be caused due to an order made under Section 11(1).

However, the Act is silent with regard to the manner in which such adverse financial impact is to be offset. There is no provision in the Act or the rules/regulations made there under which would guide the Appropriate Commission whilst offsetting an adverse financial impact.

It is stated that the Appropriate Commission may formulate regulations which could provide for the manner in which such cases are to be disposed of. This would ensure that there is uniformity in the manner in which the adverse financial impact is to be offset. These regulations could possibly deal with the principles to be adopted whilst quantifying the adverse financial impact.

Further, Section 11 of the Act could also be amended to ensure that the orders/directions issued under Section 11 are time-bound, i.e., the orders issued by the Appropriate Government are limited to a particular time period and do not last for indefinite period of time. This may be a fixed period as mentioned in the order, say 30 days or until directions are withdrawn by the concerned Government or until the extraordinary circumstances cease to persist which had justified making directions under Section 11.

It is stated that the abovementioned recommendations could to some extent ensure that the provisions of the Act are not misused and the Appropriate Government does not issue arbitrary orders which are in blatant disregard of the factual scenario and the sectoral realities.

5. Proposed Amendments

This section highlights certain amendments that could be made to sections 11, 108 and 110 of the Act incorporating the changes suggested in the previous sections.

a. Section 11

Currently Section 11 of the Act reads as follows:

'*Section 11: Directions to Generating Companies.*

(1) The Appropriate Government may specify that a generating company shall, in extraordinary circumstances operate and maintain any generating station in accordance with the directions of that Government.

Explanation: For the purposes of this section, the expression "extraordinary circumstances" means circumstances arising out of threat to security of the State, public order or a natural calamity or such other circumstances arising in the public interest.

(2) The Appropriate Commission may offset the adverse financial impact of the directions referred to in sub-section (1) on any generating company in such manner as it considers appropriate.'

Section 11 incorporating the proposed amendments reads as follows:

'Section 11. Directions to generating companies.

(1) The Appropriate Government may specify that a generating company shall, in extraordinary circumstances operate and maintain any generating station in accordance with the directions of that Government.

Explanation: For the purposes of this section, the expression "extraordinary circumstances" means circumstances arising out of threat to security of the State, public order or a natural calamity or such other circumstances arising in the public interest.

(2) Any directions of the Government under sub-section (1) shall continue to be in force for a specified period of time specified in the direction itself, or until the extraordinary circumstances under sub-section (1) cease to exist whichever is earlier.

Explanation: A specified period of time would be no longer than three months, unless for reasons specified in writing the same is extended by the Appropriate Government by a further period of one month.

(3) The Appropriate Government shall consult the Appropriate Commission in relation to any proposed legislation or rules concerning any policy direction and may take into account the recommendations made by the Commission.

(4) If any of the measures taken by the Appropriate Government under sub-section (1) has an effect of imposing restrictions on the freedom of trade, commerce or intercourse with or within the State as may be required in the public interest, then the same must be with the previous sanction of the President of India.

(5) The Appropriate Commission may offset the adverse financial impact of the directions referred to in sub-section (1) on any generating company in such manner as it considers appropriate, including through specifying Regulations governing the compensation principles as applicable from time to time.'

b. Section 110

Currently Section 110 of the Act reads as follows:

'Section 110: Establishment of Appellate Tribunal.

The Central Government shall, by notification, establish an Appellate Tribunal to be known as the Appellate Tribunal for Electricity to hear appeals against the orders of the adjudicating officer or the Appropriate Commission under this Act.'

Section 110 incorporating the proposed amendments reads as follows:

'*Section 110. Establishment of Appellate Tribunal.*

The Central Government shall, by notification, establish an Appellate Tribunal to be known as the Appellate Tribunal for Electricity to hear appeals against the orders of the adjudicating officer or the Appropriate Commission or the directions of the Appropriate Government under this Act.'

c. *Section 108*

Currently Section 108 of the Act reads as follows:

'*Section 108: Directions by State Government.*

(1) In the discharge of its functions, the State Commission shall be guided by such directions in matters of policy involving public interest as the State Government may give to it in writing.

(2) If any question arises as to whether any such direction relates to a matter of policy involving public interest, the decision of the State Government thereon shall be final.'

Section 108 incorporating the proposed amendments reads as follows:

'*Section 108: Directions by State Government.*

(1) In the discharge of its functions, the State Commission shall be guided by such directions in matters of policy involving public interest as the State Government may give to it in writing.

(2) If any question arises as to whether any such direction relates to a matter of policy involving public interest, the decision of the State Government thereon shall be final.

(3) The State Government shall consult the Appropriate Commission in relation to any proposed legislation or rules concerning any policy direction and may take into account the recommendations made by the Commission.'

This chapter contains recommendations made as a follow up to IPPAI's Legal & Tax Summit 2009.

11 Competition in Retail Supply

Anju Rani

Comparisons are often cited between the telecom and the power sector, and the general view is that power sector reforms have failed to give the sector the impetus that the telecom sector has received post its liberalisation. While it is certainly true that before the spectacular success of the telecom reforms, the power sector appeared a poor laggard, it is also true that the course charted by the power sector reforms has been a slow yet steady one: there has been substantial private participation in the generation side; there are multiple fully operational power exchanges at the national level; and significant investments are planned in the inter-state transmission networks. However, discoms continue to be the buyer of choice for most medium to large generation projects for the bankability they provide. Further, while many of the vertically integrated erstwhile state electricity boards have been unbundled and corporatised, and some discoms even privatised, the efficiencies of the private sector have yet to become visible in the day to day operations of most discoms. The conventional wisdom is that retail competition holds the key to success in the power sector.

The Electricity Act, 2003 very clearly envisages competition in the supply of electricity to consumers. In fact, a consumer may procure power directly from any of the three sellers of power—the generator, the discom and the trader. Enabling this statutorily granted choice and access for the consumer is critical for creation of a retail market for electricity.

Presently, most State Electricity Regulatory Commissions (SERCs) have notified intra-state open access regulations under which consumers requiring supply of electricity, where the maximum power to be made available at any time exceeds one megawatt, may directly enter into contracts for supply of such power with a generator or a trader, or even buy electricity at the power exchange, subject to payment of certain specified charges, including a cross-subsidy surcharge and an additional surcharge. Consumers that qualify under the aforesaid category are mostly industrial, commercial, or similarly large consumers. They have the option of using the grid to wheel such power to their premises, or, where feasible, avail such power by way of a dedicated transmission line laid and maintained by the generator connecting its plant to the consumer's installation. A detailed procedure is prescribed for availing open access through the grid—this may be for open access required to avail power purchased from outside the state (inter-state open access) or power purchased within the state (intra-state open access), and the open access required to wheel power through the distribution system of the discom, in whose area the consumer is located, up to the consumer's premises (covered under intra-state open access).

In the implementation of open access for retail consumers, several bottlenecks have been encountered. For example—a delay or non-compliance with the procedure prescribed for giving an NOC or standing clearance by the SLDC; unjustified denial of consent or denial on grounds of alleged system constraints on the part of the discom; lacunae in the regulatory framework such as a delay in or failure to determine/prescribe applicable open access charges such as reactive energy charges; lack of clarity on UI issues and ABT settlement mechanisms; and uncertainty on issues of continuity of supply on system/scheduled outages.

Fears of loss of revenue by the incumbent discom and cherry picking by competing sellers are expected apprehensions and issues, as is evident in reforms carried out in other countries; for example during the course of the electricity reforms in the UK on whose model our own Act is supposed to be based. Further, it is noteworthy that the Utilities Act, 2000 of UK prescribes separate entities for distribution and supply at the retail level. Thus, under the Act, a distribution licensee's function is to build and maintain the distribution network and provide connections to new consumers; whereas the supply companies are required to undertake supply to the consumers and may compete with other suppliers to do so. In this respect, there is a significant structural difference between the Indian model and the UK model. In the Indian model, a distribution licensee is licensed to distribute as well as supply, and thus carries a stronger monopolistic feature.

Competition, by definition, requires competing interests. However, the benefits of competition, including optimisation of resources, are only possible if a level playing field is provided to the competing interests, such that good performance is incentivised and poor performance disincentivised. It is also noteworthy that, institutional and procedural barriers that impede access may also take the form of time and costs involved in remedying a wrong. For example, the onus on the open access consumer to seek regulatory intervention for resolving issues in procuring open access may work with retail consumers with structured organisations and resources; however it would be a deterrent, going forward, when consumers with demand lower than one megawatt become eligible to seek supply through open access.

Recourse to adjudication must always remain an option; however there is a persuasive case for a more active regulatory role and oversight in respect of open access, especially at the initial stages of creation of retail electricity markets. While care has been taken to design open access regulations that are enforceable by providing timelines, requiring reasoned decision-making and prescribing online, publicly available, status reports on open access by SLDCs, in case of non-compliance, the aggrieved thus far has had to petition the regulator for a remedy—a process that may be short, when compared to general litigation proceedings, but still a potential deterrent for retail open access consumers. On the other hand, setting up a dedicated open access monitoring cell under the SERC secretariat, that sources and maintains the open access details with pendencies, reasons for non-availability and other relevant data, and undertakes regular analysis and audits of the same with recommendations, would appear to be a more dynamic and efficient option. More than anything else, this will serve as a step towards an institutionalised pro-access environment. This will not only be of assistance in quick identification of problem areas but will also complement the process of system planning, strengthening and expansion, on an ongoing basis. Further, a strict approach to require adherence to the standards of performance may also serve the need of the hour.

There is also a persuasive case for tweaking regulations to take into account the realities of the varying nature of 'premises' in existence, and introduce targeted amendments to enable consumers and persons residing or occupying buildings and areas such as multi-storied buildings/high-rises, housing colonies, commercial complexes and industrial parks, to avail open access for sourcing electricity from competing suppliers. The distinction made in the Act between the terms 'distribution' and 'supply' has hampered the occupiers of the aforesaid buildings and areas from availing electricity through open access. While the building owner or an association of persons or individuals, or society constituted by the occupiers may contract with a generator or trader for electricity in bulk (and at a single point) and the generator or trader is permitted by the Act to 'supply' electricity to such open access consumers, ambiguity surrounds the issue of sharing/collection of the charges and costs, including sharing of transformational losses and losses in the internal distribution system, by the owner/association/society from the tenants/constituents/members thereof, as such an act may be tantamount to sale and/or distribution of electricity, which may not be undertaken without a licence. Thus, an enabling provision under the regulations that recognises the above consumer groups as a separate consumer category and exempts the collection/sharing of charges and losses from licence requirement, or clarifies the internal distribution activity as not constituting 'distribution', along with prescribing adequate safeguards to protect consumer interests, would make open access a viable option for many residents and consumers. It is no secret that there are many such consumer groups who are currently paying astronomical rates for reliable, round the clock power, generated through onsite diesel gensets which are polluting, besides clearly being the less economical option to grid power. Generating stations should be enabled to tap this market. Channelling funds to more efficient forms of generation will surely be in the best interests of all stakeholders.

Further, the SERC may also consider liberalising the one megawatt limit to open up the retail market to a larger number of consumers. This may also assist in structuring contracts that do not necessarily require an association or a bulk buyer for an individual consumer to be able to source electricity through open access.

Finally, with the mandate being greater investment in and creation of new and good quality infrastructure, the time may have come to pro-actively ramp up capacity in the distribution systems, so that system constraints are no longer an easily acceptable ground for denial of open access. For the Act to be realised in letter and spirit, it is imperative that the pure wires business at the distribution level is no longer perceived as the least remunerative business proposition. The future appears to point towards partnerships needed for deployment of sophisticated technology solutions at the distribution level—be it the smart grid or bringing each household as a stakeholder and beneficiary into the energy efficiency market.

12 Functioning of Electricity Exchanges and Markets

Rupa Devi Singh

Introduction

Competition, as the word suggests, is the struggle between rivals to attain a certain objective. Competition in any sphere of activity leads to higher degree of inputs and efforts which results in higher degree of efficiency. The presence of competition in any industry implies the presence of rivalry among suppliers of the said commodity or service. Different strategic actions are taken by firms competing in a market to attract consumers.

Economies across the globe are looking at reaping the benefits of an open market place where the consumers are the ultimate winners. One of the key objectives of electricity reforms initiated in India in 1991 was to promote fair competition in the electricity sector. Steps were taken to convert the traditionally monopolistic system into a competitive and market oriented industry. The Electricity Act, 2003 was a major step in this regard and superseded all concerned previous Acts such as Indian Electricity Act, 1910, Electricity Supply Act, 1948 and Electricity Regulatory Commissions Act, 1998.

For an increased consumer welfare competition needs to ensure a fair market place for all its participants. A level playing field where, the retail businesses, distribution companies or retail agents can compete with each other to attract the ultimate users of service. This would ensure cost-cutting, increase in efficiency and innovation which would in turn lead to expansion of output as well as reduction in price; improvement in the quality of service; as well as wider options in choosing suppliers.

One of the key objectives of Indian electricity reforms has been to promote healthy competition in the power market in the country. Unrestricted open access to the transmission and distribution network has been recognised as a key driver of competition in wholesale as well as retail markets.

The Central Electricity Regulatory Commission (CERC) has passed the required regulations and taken many steps to promote competition in this market. However, state level reforms have not fully addressed the issue of fair access to the common carrier. Significant captive capacity remains unutilised when it could have been injected into the grid.

Policy and Regulatory Reforms

As envisaged in the Electricity Act, 2003 as well as National Electricity Policy, 2005 (NEP), unobstructed open access is a key requirement for facilitating competition in wholesale as well as retail electricity markets. The Act authorises respective regulatory commissions to specify various norms and charges, including cross-subsidy surcharges, for availing of open access. Generation companies may sell energy to any potential buyer in the country and vice-versa.

However, despite some key initiatives such as enforcement of open access regulations and constitution of power exchanges on part of the CERC, the status of open access across various states is poor and thus effective competition is lacking.

The Indian power market have been historically illiquid in nature, with only approximately 8 per cent of electricity produced being traded through short-term market. However the exchanges have proved to be vital in transparent price discovery. The prices discovered on the exchanges have been well accepted by the market and are beginning to be used as reference prices for other transactions in the market.

Trading in the power markets has witnessed steady growth over time. In volume terms (kWh) the size of the short-term market in India, excluding UI and direct bilateral sale between distribution companies, was about 30.6 billion kWh (units) in the year 2009. As compared to the volume of electricity transacted through short-term market in the year 2008, this was about 20.3 per cent higher.

During the Month of August 2010, total electricity generation excluding generation from renewable and captive power plants in India was 67174.98 of which around 11.86 per cent was transacted through short-term contracts.

Whereas during the month of January 2011, total electricity generation excluding generation from renewable and captive power plants in India was 71282.05 MUs. Of the total electricity generation in January 2011, 7229.65 MUs (10.14 per cent) were transacted through short term, comprising 3274.18 MUs (4.59 per cent) through bilateral (through traders and term-ahead contracts on power exchanges and directly between distribution companies), followed by 2527.68 MUs (3.55 per cent) through UI and 1427.79 MUs (2.00 per cent) through day ahead collective transactions on power exchanges (IEX and PXIL).

The power exchanges (PXs) started their operations with participation from utilities and subsequently gained participation from various generating companies, captive plants and industrial participants. The exchanges have currently more than 900 registered participants and the numbers are only increasing. The advent of intra-state open access further made participation of smaller IPPs a reality. Now, with the regulations being in place and a number of reforms being aimed at improving the market participation the awareness about trading on the exchanges has also witnessed an uptrend.

Open access had made it possible for smaller retail consumers to procure power on a short-term basis. Further the availability of power at competitive price at PXs provides uninterrupted production to the industries.

The national platform provided by the PXs has been able to match short-term surplus with demand variation in diverse parts of the country. The exchanges have provided an equal bargaining power to all stakeholders in a supply deficit environment. Moreover, an assured payment structure and a prudent risk system have further led to increase in participation in the markets.

However, the percentage of trades through the power exchanges have only been a small part of the total power production in the country.

An unrestricted open access is not only necessary for market competition but also for attracting potential investors in this segment. Constraints on open access take various forms. In several states, captive power plant owners and independent power producers are offered very low tariffs by domestic distribution companies. Also, a private generating unit interested in selling energy out of the state is not provided the needed access to the transmission system. A huge chunk of capacity remains unutilised in many states such.

Another important driver of competition in the distribution sector is parallel licensing, which has not taken off either. Regulatory bodies are mandated to issue parallel licences in the distribution segments to facilitate competition in retail markets. However, till date, Mumbai is the only supply area where parallel licences have been issued: two companies—Reliance and Tata Power—which are competing with each other. Majority of the other areas are characterised by monopoly status of utilities.

In most of the states, the progress on further reforms in distribution segments is very slow. On the other hand, an increasing trend has been observed in the tariff for power traded in the short-term market due to capacity shortages and increasing demand at the consumer end. The peak price is reported to be two-three times the normal rate of power purchase. High electricity purchase prices have put additional burden on electricity consumers. Various stakeholders have started raising concerns over the poor status of competition in the sector.

The current structure doesn't allow a producer a fair opportunity to sell power in the state outside its state of production. This implies that the system lacks fair and level playing field for competition. As required by the Electricity Act, 2003 as well as National Electricity Policy (2005), some steps have been taken to ensure unbarred open access at the state level. Planning Commission has proposed that at least 10 per cent of the total available power should be sold through open access.

In order to facilitate intra-state trading of power, about 21 SERCs have formally enforced open access regulations. CERC has also issued the regulations for inter-state transactions. These regulations mainly require transmission as well as distribution companies to allow open access by generating companies as well as consumers with a connected load in excess of a megawatt (MW).

There is substantial progress in formally enforcing regulations on open access, out of the select 23 states where electricity regulatory commissions are fully functional, 21 states have formally enforced the open access regulation. Similarly, there is adequate progress in specifying surcharges and wheeling charges. Responding to the mandate to ensure open access for large consumers, 19 states have notified that consumers with connected load in excess of 1 MW may apply for open access.

Power purchased from major sources such as electricity generating companies is referred to as bulk supply business. Earlier, contrary to provisions of the Act, state transmission utilities (STUs) were also functioning as bulk supply licensees. To ensure independence of the common carrier business, STUs are prohibited from engaging in bulk supply business. This has resulted in a separation of transmission utility from the bulk supply business.

The status of open access is in a very poor state. Inspite of a provision being made to allow distribution companies as well as consumers to use transmission facilities by paying wheeling charges to STUs, in practice implementation is lacking. Open access, in its truest sense, is almost absent due to transmission capacity shortages and poor response from distribution companies to enforced regulations.

Open Access in the true sense of the word is still not a reality. Reluctance on part of certain utilities and inconsiderate responses have led to a huge failure in reaching the retail consumers.

Most of the states in the country are facing acute power shortages. Because of the failure to meet targets for capacity addition during the last two five year plan periods, the gap between demand and supply of power has widened.

Given this acute shortage, various state governments are not allowing private players, especially captive power plants, to export energy to other states. On the other hand, a few utilities source power from captive plants at low rates and sell it at attractive prices to other states.

Section 11(1) which states that 'generating company shall in extraordinary circumstances operate and maintain any generating station in accordance with the directions of that Government', based on this, some state governments have banned outside sale of power by domestic power producers. Whereas, power shortages prevailing throughout the year cannot be justified as 'extraordinary circumstances'.

Regulatory steps need to be taken to ensure elimination of such state induced barriers to open access.

Apart from the reluctance on part of certain utilities and inconsiderate responses, the failures in open access have been attributed to unavailability of transmission capacity. The transmission capacity available with Central transmission utility (CTU) as well as STUs is not adequate for meeting the increasing demand for open access, especially for inter-state power transactions.

Certain transmission utilities do not operate independently. Some of them are controlled by the state government management and hence have to follow the directions provided to them. The Electricity Act, 2003 as well as National Electricity Policy, 2005 sought to facilitate independent functioning of transmission utilities and load despatch centres to ensure competitive neutrality in the power business. This has not been entirely possible in the current state of affairs where the control, directly or indirectly lies with the state government.

Mumbai is a good example of consumers being provided a choice in selecting suppliers. Open access in the distribution sector should allow for the same distribution network to be utilised by two or more companies to serve the consumers in a specific area. Mumbai has provided a very successful example of such open access. Other ERCs should also be encouraged to try and replicate this model. The model can be easily replicated in territories where energy availability is reasonably good and the transmission and distribution network is adequate.

Provisions in the Act requiring regulatory bodies and utilities to ensure fair and non-discriminatory open access to all players in the market are adequate. The Electricity Act, 2003 as well as National Electricity Policy, 2005 has tried to promote competition in the power sector by facilitating open access in wholesale as well as retail markets.

On the same lines, CERC as well as respective SERCs have issued regulations to facilitate inter-state and intra-state open access respectively and have specified norms and surcharges as well as wheeling charges to be paid to distribution companies for availing of open access.

The progress in this sector has been noteworthy, the progress in notifying the required regulations and guidelines has been well in lines with the intent of the Act. However, the status of open access is not very satisfactory across states.

Discriminatory actions taken by some state distribution companies/ SEBs have adversely affected open access; the volume of energy transferred through open access is almost negligible.

Lack of open access has adverse impacts on the markets: distribution companies in respective areas continue to enjoy monopoly; consumers are exploited due to lack of choices, the generation business suffers from the lack of incentive for investment due to inadequate access to transmission systems across geographies, etc. At the same time, a few distribution companies are exploiting the mentioned shortages and charging very high prices for short-term sale across states. These are against the spirit of the law and policy.

Given the above discussion relating to the need for open access and its poor facilitation at present, a few stringent steps need to be taken to ensure adherence to the Electricity Act, 2003. The implementation of the Act in spirit and in letter needs to be ensured. One of the biggest constraints is transmission capacity, which needs its due share of consideration and investments.

The existing regulations may be reviewed at frequent intervals to identify and eliminate any hurdles that are acting as impediments in the implementation of the Act. More autonomy should be provided to the STUs and load despatch centres to ensure competitive neutrality and provide for 'independence' in its true sense. Wheeling charges and cross-subsidy surcharges should be low enough to not have a significant adverse impact on the demand for open access. Another major area of impact, which is often ignored, is the need for adequate capacity building; discussions should be initiated through conferences/seminars to remove misconceptions about open access. It is necessary to increase awareness among state governments about the long-term benefits to the industry as well as consumers from open access.

Given the growth that the open access regime has promised and the fast growing acceptability of the trading mechanism provided by the power exchanges, the power markets have been aligned to a growth path. An increasing number of participants and the growth in trading volumes over the exchanges signify the acceptance of this mechanism by the market participants. Proper regulatory impetus and free and open access to the transmission systems together can help create a transparent and trustworthy trading environment for the consumers.

References

Barpanda, S.S. (2010). "Open Access in Inter State Transmission System", 3rd Capacity Building Programme for Officers of Electricity Regulatory Commissions, IIT Kanpur, 23-28 August. DGM, GM, NRLDC.

Kumar, Rajesh (2010). "Promoting Competition though Open Access in the Power Sector", *Briefing Paper* No. 4. Jaipur: CUTS International.

13

Captive Power-Policy and Regulatory Framework

Balawant Joshi

Background

Electricity is an essential input to the economy and an efficient and competitive power sector is vital to a country's development. Commencing with a meagre installed capacity of around 1,350 MW in 1947, the Indian power sector has made substantial progress over the last six decades and the installed capacity at the end of Ninth Five Year Plan (1997-2002) increased to 1,05,000 MW.

Figure 13.1

Growth of Electricity Generation Capacity

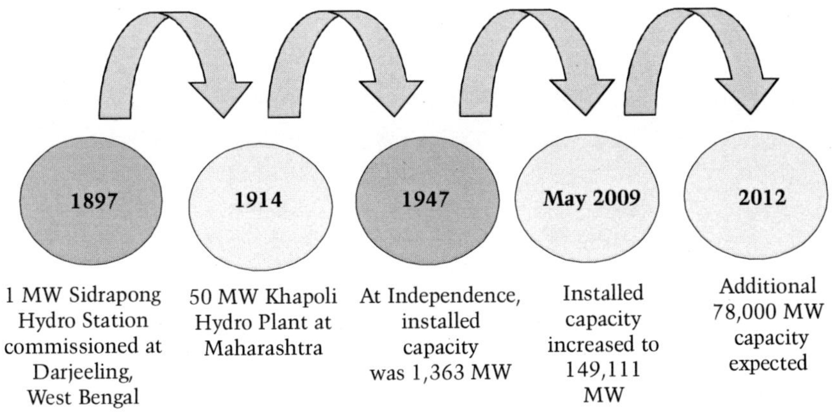

However, the installed generation capacity is inadequate to meet the increasing demand. Currently the country is facing energy shortage of around 11 per cent and peak shortage of about 12 per cent. The demand supply position over the last seven years and energy shortfall during the last seven years is as follows:

Figure 13.2

Peak Demand Supply Position

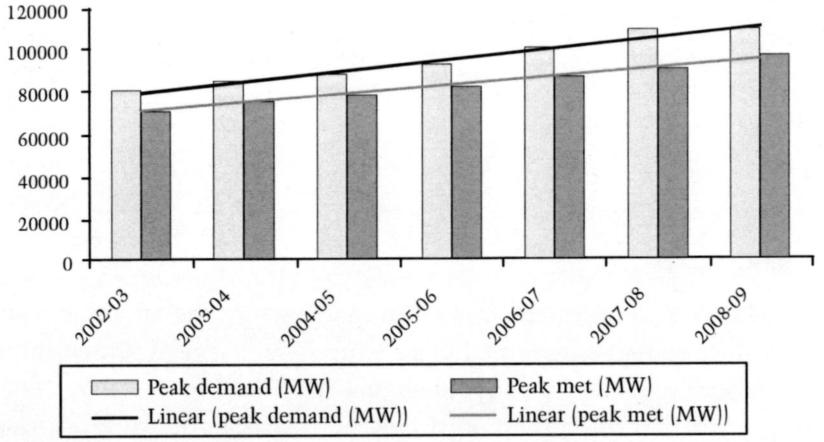

Source: CEA and ABPS Research, 2010.

Figure 13.3

Energy Requirement and Availability

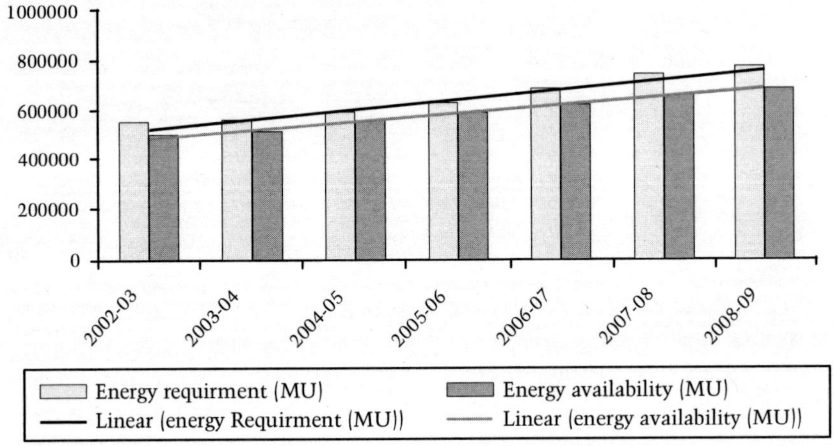

Source: CEA and ABPS Research, 2010.

Nearly all states, including the most industrialised like Maharashtra, Karnataka, Andhra Pradesh, Tamil Nadu, etc., are reeling under acute power shortage during peak hours. The distribution utilities are adopting load shedding measures for bridging the demand-supply gap and presently, the load shedding measures are limited not only to the towns and villages, but metropolitan and urban areas are also facing load shedding of 2-6 hours varying from state to state. Industrial production is also adversely affected due to load shedding, mandatory weekly holiday etc. In order to alleviate the plight of consumers from load shedding and frequent power cuts, several models are being tried out in the country. The distribution utilities are procuring power from electricity traders, cogeneration and captive power plants, renewable energy sources, etc. Thus, in the current power shortage scenario, the captive power plants (CPPs) are not only meeting their own requirements but also are helping in reducing the load shedding to other consumers.

Overview of Captive Power in the Country

Industrial sector is one of the largest consumers of electrical energy in India. However, a number of industries over a period of time are relying on their own generation (captive power generation and/or cogeneration) rather than depending on the grid supply, primarily due to the following reasons:

- Non-availability of adequate grid supply (huge demand supply shortage).
- Poor quality and reliability of grid supply.
- High tariff as a result of heavy cross-subsidisation.

CPPs are essentially non-utility power plants and are owned by specific industry(ies) which consume majority of the power produced in their production operations. The total installed captive power capacity in the country is around 32,000 MW.

Sector-wise Captive Power Plants

The sector-wise installed captive power capacity is given in the Figure 13.4.

As observed from Figure 13.4, around 33 per cent of the total captive power capacity has been installed in metals and minerals industry followed by 12 per cent in chemicals, 11 per cent in cement, 10 per cent in engineering, 9 per cent in textile and remaining in other sectors such as sugar, petrochemicals, etc.

Figure 13.4

Sector-wise Installed Captive Power Capacity

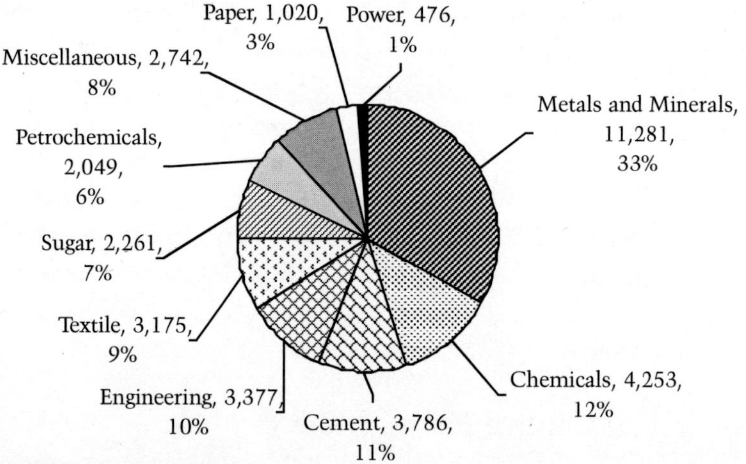

Source: *India Infrastructure Report, 2009.*

CPPs: Capacity Ranges

In terms of capacity, the size of CPPs varies significantly. While some of the CPPs are as small as 0.2 MW, a few others are of more than 300 MW in size. The number of CPPs along with capacity in various range of size is given in Figure 13.5.

Figure 13.5

No. of CPPs and Corresponding Installed Capacities in Various Size Ranges

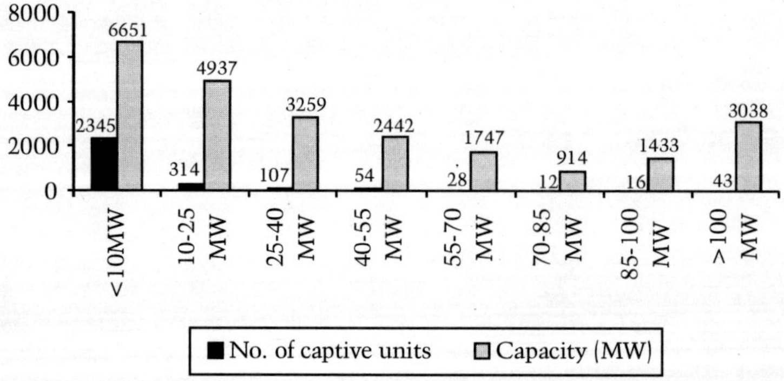

Source: *India Infrastructure Report, 2009.*

As observed from the above figure, the maximum numbers of CPPs i.e., 2,345 are of less than 10 MW and installed capacity of these CPPs is 6,651 which corresponds to average size of around 3 MW each.

CPPs: State-wise

The state-wise installed capacity of CPPs in the country is pictorially depicted Figure in 13.6.

Figure 13.6

State-wise CPPs

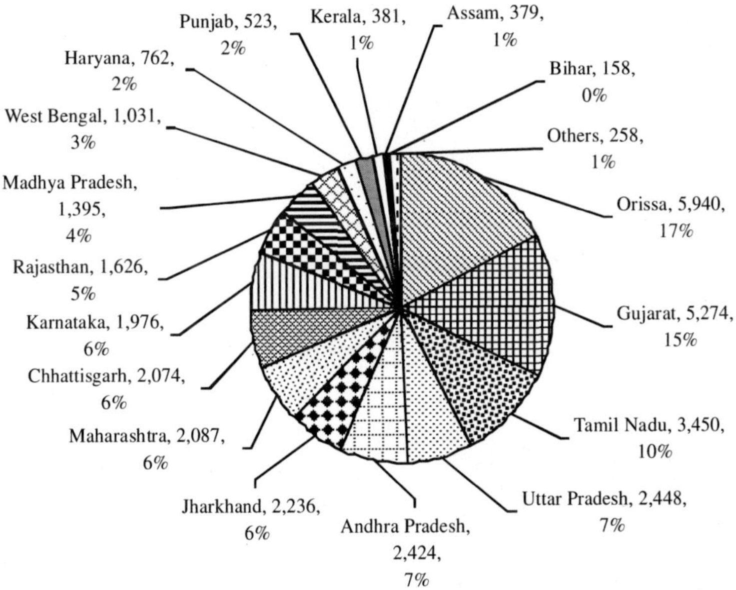

CPPs: Fuel/Source-wise

Coal is the major fuel source for CPPs in India and constitutes around 42 per cent of total 32,000 MW captive capacity, followed by liquid fuels which constitutes around 24 per cent of total captive capacity. Figure 13.7 below demonstrates the fuel-wise installed capacity of captive power in India:

Figure 13.7

Fuel-wise CPPs

Fuel-wise Installed Capacity (MW)

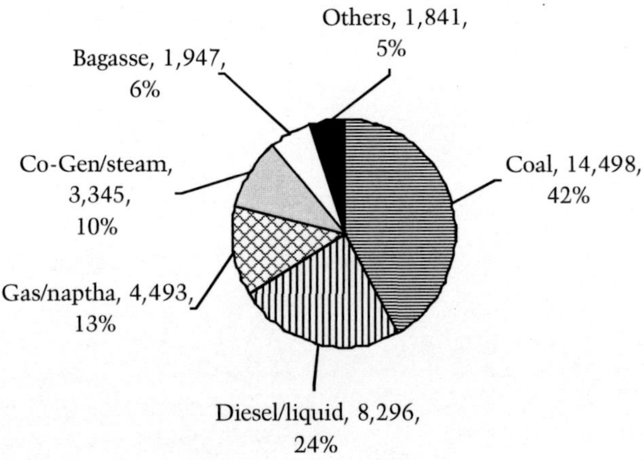

Source: India Infrastructure Report.

In 2005, the total captive power generation using coal as fuel constituted 44 per cent of total generation, while the same has been reduced to 42.1 per cent in 2008 and is expected to be reduced further with an expected increase of wind and other renewable sources.

Policy and Regulatory Framework for Captive Power Generation

Prior to enactment of Electricity Act, 2003, industrial sector in many states was at the receiving end in terms of supply of electricity. Unscheduled power cuts, fluctuating voltage and frequency, resulted in huge production and revenue losses to the industry. In order to guard themselves against such erratic supply, industries explored various options including setting up their own generating plants mainly for their own use. Repealed Indian Electricity Act, 1910 and Electricity (Supply) Act, 1948 had no specific provisions for construction, operation of captive generating plants, and the development of captive generating plants was essentially dependent on the CPP policies formulated by the respective state governments. Also, development of CPP was difficult as it entailed availing of many clearances and approvals from Central Electricity Authority, state government, and State Electricity Board in accordance with the policies prevalent in respective states for setting up captive generating stations.

However, pursuant to enactment of Electricity Act, 2003 and subsequent notifications of the National Electricity Policy (NEP) and Tariff Policy (TP) by the Central government, the issue of clarity and uniformity in policy and regulatory framework has been specifically addressed in respect of CPPs. This has facilitated their development. The provisions related to captive generating plants as mentioned in Electricity Act, 2003, NEP and TP with their implications are discussed in the following paragraphs.

Provisions in Electricity Act, 2003

Section 9 of Electricity Act, 2003 contains the provisions related to construction of captive generating station and construction of dedicated transmission lines, and is reproduced below:

> 9. (1) Notwithstanding anything contained in this Act, a person may construct, maintain or operate a captive generating plant and dedicated transmission lines:

Provided that the supply of electricity from the captive generating plant through the grid shall be regulated in the same manner as the generating station of a generating company.

Electricity Act 2003 has provided the definition of captive generating plant under Section 2(8) and the same is reproduced below:

> "Captive generating plant" means a power plant set up by any person to generate electricity primarily for his own use and includes a power plant set up by any co-operative society or association of persons for generating electricity primarily for use of members of such cooperative society or association;

From the above definition, it can be concluded that captive generating station can be set up by any person, cooperative society and association of persons (AoP), primarily for own use.

Person

In this regard, the definition of 'person' assumes great significance. The definition of person as provided in the Section 2(47) of Electricity Act, 2003 is reproduced below:

> "person" shall include any company or body corporate or association or body of individuals, whether incorporated or not, or artificial juridical person;

Electricity Rules, 2005

Although, Electricity Act, 2003 has provided the basic framework for establishing the captive generating plant, there was ambiguity over the 'eligibility criteria' and interpretation of the phrase 'primarily for own use' for the purpose of meeting requirements of captive generating plant. To clarify this matter, Ministry of Power, Government of India (GoI), notified a set of rules namely 'Electricity Rules, 2005' on 8 June 2005.

Eligibility Criteria for Captive Generating Plant

The Clause 3 of Electricity Rules, 2005 has specified the minimum eligibility criteria in terms of energy consumption as well as ownership stake in the captive generating station. The relevant text of the Clause 3 of Electricity Rules, 2005 is reproduced below:

(1) No power plant shall qualify as a "captive generating plant" under section 9 read with clause (8) of section 2 of the Act unless—

(a) in case of a power plant:

  (i) not less than twenty six percent of the ownership is held by the captive user(s), and

  (ii) not less than fifty one percent of the aggregate electricity generated in such plant, determined on an annual basis, is consumed for the captive use.

Therefore, any person must have at least 26 per cent equity stake in the power plant and must have at least 51 per cent energy consumption on annual basis out of total electricity generated by the power plant to qualify it as CPP.

Further the 'captive user' in Electricity Rules, 2005 is defined as end-user of the electricity generated in captive generating plant and the term 'captive use' shall be construed accordingly.

Captive Generating Plant by AOP or Cooperative Society

Further, the Clause 3 of Electricity Rules, 2005 has elaborated on the eligibility criteria for qualification of CPP, in case such plant is being set up by an AoP, or a cooperative society. The relevant text of Clause 3 is reproduced below:

> Provided that in case of power plant set up by registered cooperative society, the conditions mentioned under paragraphs at (i) and (ii) above shall be satisfied collectively by the members of the co-operative society:
>
> Provided further that in case of association of persons, the captive user(s) shall hold not less than twenty six percent of the ownership of the plant in aggregate and such captive user(s) shall consume not less than fifty one percent of the electricity generated, determined on an annual basis, in proportion to their shares in ownership of the power plant within a variation not exceeding ten percent;

Here, the point to be noted that energy consumption by the members will have to be in proportion of their equity stake in the captive generation plant. Variation in the proportionate consumption up to 10 per cent has been allowed.

Captive Generating Plant by Special Purpose Vehicle

Further, the Clause 3 of Electricity Rules, 2005 has elaborated on the eligibility criteria for qualification of CPP, in case such plant is being set up by special purpose vehicle. The relevant text of Clause 3(b) is reproduced below:

> In case of a generating station owned by a company formed as special purpose vehicle for such generating station, a unit or units of such generating station identified for captive use and not the entire generating station satisfy (s) the conditions contained in paragraphs (i) and (ii) of sub-clause (a) above including—

Explanation:-

(1) The electricity required to be consumed by captive users shall be determined with reference to such generating unit or units in aggregate identified for captive use and not with reference to generating station as a whole.

(2) The equity shares to be held by the captive user(s) in the generating station shall not be less than twenty six per cent of the proportionate of the equity of the company related to the generating unit or units identified as the captive generating plant.

(3) It shall be the obligation of the captive users to ensure that the consumption by the Captive Users at the percentages mentioned in sub-clauses (a) and (b) of sub-rule (1) above is maintained and in case the minimum percentage of captive use is not complied with in any year, the entire electricity generated shall be treated as if it is a supply of electricity by a generating company.

Further, the special purpose vehicle (SPV) is defined as 'Special Purpose Vehicle shall mean a legal entity owning, operating and maintaining a generating station and with no other business or activity to be engaged in by the legal entity.'

Thus the rules restrict the SPV created for developing, operating and maintaining CPP from engaging in any other activity.

Statutory Provisions under NEP and TP

The National Electricity Policy (NEP) highlights that the GoI has initiated several reform processes to create a favourable environment for addition of new generating capacity in India. Thus, the captive generation has been freed from all controls like licence, techno-economic clearance from CEA etc.

NEP, in regard to captive power generation specifies as follows:

> The liberal provision in the Electricity Act, 2003 with respect to setting up of captive power plant has been made with a view to not only securing reliable, quality and cost effective power but also to facilitate creation of employment opportunities through speedy and efficient growth of industry.

> The provision relating to captive power plants to be set up by group of consumers is primarily aimed at enabling small and medium industries or other consumers that may not individually be in a position to set up plant of optimal size in a cost effective manner. It needs to be noted that efficient expansion of small and medium industries across the country would lead to creation of enormous employment opportunities.
>
> A large number of captive and standby generating stations in India have surplus capacity that could be supplied to the grid continuously or during certain time periods. These plants offer a sizeable and potentially competitive capacity that could be harnessed for meeting demand for power. Under the Act, captive generators have access to licensees and would get access to consumers who are allowed open access. Grid inter-connections for captive generators shall be facilitated as per section 30 of the Act. This should be done on priority basis to enable captive generation to become available as distributed generation along the grid. Towards this end, non-conventional energy sources including cogeneration could also play a role. Appropriate commercial arrangements would need to be instituted between licensees and the captive generators for harnessing of spare capacity energy from captive power plants. The appropriate Regulatory Commission shall exercise regulatory oversight on such commercial arrangements between captive generators and licensees and determine tariffs when a licensee is the off-taker of power from captive plant.

The Tariff Policy (TP) notified by Ministry of Power recognises that the state governments have the rights to impose duties, taxes, cess on generation and consumption of electricity from CPPs. However, it has advised state governments to avoid discriminatory treatment meted out to CPPs by levy of excessive taxes thereby rendering the consumption from CPPs uneconomical. The relevant extract of the TP provision is as under:

> In some cases, the duties etc. on consumption of electricity is linked to sources of generation (like captive generation) and the level of duties levied is much higher as compared to that being levied on the same category of consumers who draw power from grid. Such a distinction is invidious and inappropriate. The sole purpose of freely

allowing captive generation is to enable industries to access reliable, quality and cost effective power. Particularly, the provisions relating to captive power plants which can be set up by group of consumers has been brought in recognition of the fact that efficient expansion of small and medium industries across the country will lead to faster economic growth and creation of larger employment opportunities.

TP further entrusts responsibility of crating an enabling environment on the Appropriate Commission. Relevant extract of the TP states:

> Captive generation is an important means to making competitive power available. Appropriate Commission should create an enabling environment that encourages captive power plants to be connected to the grid.

Electricity (Amendment) Act, 2007

While the Electricity Act, 2003, NEP and TP provided ample clarity on the issue of setting up of captive generating plant, there was ambiguity on the issue of sale of surplus power from the CPP to licensees or to open access consumer. The Hon. Supreme Court in the matter of captive power plant case of M/s Bhushan Steel Ltd. in Maharashtra had upheld the judgment of Maharashtra High Court in WP No.882/05 (MSEB *Vs* State of Maharashtra & Others) that a captive generator would require a licence to sell its surplus electricity.

If the CPP would have required to seek licence from regulator to be able to sell surplus power to any other consumer, it would have rendered the entire concept of captive power ineffective. Therefore to address this specific aspect, the Government of India enacted the Electricity (Amendment) Act, 2007 which came into effect on June 12, 2007. The relevant extract (Clause 3) of Electricity (Amendment) Act, 2007 is as under:

In Section 9 of the Principal Act, in sub-Section (1), after proviso, the following proviso shall be inserted, namely:

> Provided further that no license shall be required under this Act for supply of electricity generated from a captive generating plant to any licensee in accordance with the provisions of this Act and the rules and the regulations made thereunder and to any consumer, subject

to the regulations made under Sub-Section (2) of Section 42 of the Act.

In view of above, the sale of surplus power from captive power plant to licensees or open access consumers has been enabled without the need to avail any licence for the same.

Case Study on Eligibility Criteria for SPV: Malanpur Captive Power Limited

A case of CPP set up in Malanpur, (Bhind District) Madhya Pradesh provides a reference case for establishing the group captive scheme, wherein ownership issues have been addressed pursuant to notification of MoP Rules. Madhya Pradesh Electricity Regulatory Commission (MPERC) approved the Group Captive Scheme through its Order dated May 22, 2007 (Petition No. 2/2007).

The summary of scheme and MPERC Order has been presented in the following paragraphs:

Captive power plant (CPP) of 26.19 MW has been established by three user shareholders namely SRF, Montage and Supreme and two sponsor shareholders namely Wartsila India Ltd. and Crompton Greaves Ltd. through the formation of a special purpose vehicle 'Malanpur Captive Power Limited'(petitioner) which owns, operates and maintains the generating station, has no other business or activity. CPP users are expected to consume a minimum of 13.90 MW, which translates to 69.5 per cent of the available capacity (fuel is available only for 20 MW). In this case, respondent was MP Madhya Kshetra Vidyut Vitaran Co. Ltd. Bhopal. The relevant extract of the MPERC Order is reproduced below:

> During the course of hearing on 4th May 2007, it was submitted by the respondent that various companies have become shareholders in the petitioner company and wanted to avail power supply. The petitioner company is a different entity and is not using the power for itself but is supplying/selling power to other companies for their use in their respective plants. It was submitted that Rules cannot supersede the Act. The petitioner company is nothing but a distribution company and wants to do business of sale of power under the mask of captive generation to save itself from the clutches

of obtaining a license and to deprive the respondents of payment of charges.

......

Going through the records submitted by the petitioner, the Commission is in agreement with the petitioner company that it is a captive generating plant within the meaning of Section 2(8) and section 9 of the Act. Also the petitioner's company is a "Special Purpose Vehicle" within the meaning of rule 3(1) (b) of the Electricity Rules 2005. As submitted by the petitioner, it is engaged only in the generation of power and in no other business. The petitioner has denied the allegation of the respondent that under cover of captive power plant the petitioner is selling power to other consumers. It is worth-mentioning here that the respondent has not advanced specific evidence to substantiate the allegation that the petitioner is indulging in sale of power to consumers other than CPP users.

.....

The Commission is not in agreement with the argument of the respondent that he is entitled to recover the cross subsidy surcharge as per provisions of Section 42(2) of the Act. It is provided in the 4th proviso of Section 42(2) that such charge shall not be leviable in case open access is provided to a person who has established a captive generation plant for carrying the electricity to the destination of his own use.

Various directives issued by MPERC to said SPV for implementation of Group Captive Scheme are as under:

(1) The petitioner shall strictly abide by the provisions of the Electricity Act 2003 and the Rules made there under and the regulations framed by the Commission along with the necessary amendments to be notified in the Act, rules and regulations from time to time.

(2) The petitioner shall always adhere to the following:

   (a) Not less than 26 per cent of its ownership shall be held by its CPP users holding voting rights.

(b) Not less than 51 per cent of the aggregate electricity generated in its plant, determined on annual basis shall be consumed by its CPP Users.

The respondent may bring to the notice of the Commission if any violation is noticed by them in this regard for appropriate action by the Commission.

(3) The petitioner shall give an undertaking to the respondent that the petitioner company is not engaged and shall not be engaged in other business or activity except the captive generation of power and supply as per explanation 1 (d) of rule 3 of the Electricity Rules, 2005. The respondent may bring to the notice of the Commission if any violation in this regard is noticed by them for taking appropriate action by the Commission.

(4) The petitioner shall submit to the respondent and the Commission the annual information of the status of ownership and of consumption by the captive users in the above CPP. The petitioner shall also be informing the respondent and the Commission of any changes occurring in the shareholdings and/or consumption by the shareholders within a period of one week of occurrence of such change. The petitioner has to inform the respondent immediately after commissioning of the remaining two units and status of equity pattern at the time of commissioning.

(5) According to the rule 3(2) of Electricity Act, 2005, it shall be the obligation of the captive users to ensure that the consumption by the captive users at the percentage mentioned in sub-clause (a) and (b) of sub-rule (1) above is maintained and in case percentage of captive use is not complied within any financial year, the entire electricity generated shall be treated as if it is a supply of electricity by a generating company.

(7) The CPP shall be established, operated and maintained as per technical safety and grid standards specified by the Authority or State Transmission Utility (STU) or the Commission and shall also comply with the provisions of MPERC (power purchase and other matters with respect to conventional fuel based captive power plants) Regulations, 2006.

(8) The above approval is subject to the condition that the petitioner shall obtain the written permission from GoMP for 'Right of the way' for erection of own dedicated H.T. line and permission from Chief Electrical Inspector, GoMP for erection, charging and observing safety rules.

Regulatory and Policy Framework for Transmission of Captive Power

Earlier, captive generating stations were set up within the industrial premises wherein load/consumption was situated. Hence, there was rarely any requirement of establishing elaborate evacuation and transmission arrangement with connectivity to the grid. Most of the CPP installations were on stand alone basis with 'island mode' of operation, not connected with the grid. However, pursuant to enactment of the Electricity Act, 2003, many captive power plants are being set up outside the industry premises, mainly due to space constraint within industry premises and availability of fuel sources at pit head. In order to optimise capital investment and resource utilisation, a concept of group captive power plants is also emerging wherein CPP capacities can be optimally designed and sized to meet collective load requirement in optimal and efficient manner. The Electricity Act, 2003 has provided following two mechanisms under which electricity generated from CPP can be transmitted from the remotely located CPP to its point of use:

1. Dedicated transmission lines,
2. Open access.

Wheeling of Power using Dedicated Transmission Lines

The Electricity Act, 2003 has provided the definition of Dedicated Transmission Lines under Section 2 (16) and same is reproduced below:

> "Dedicated Transmission Lines" means any electric supply line for point to point transmission which are required for the purpose of connecting electric lines or electric plants of a captive generating plant referred to in section 9 or generating station referred to in section 10 to any transmission lines or sub-stations or generating stations or the load centre, as the case may be;

The option of establishing a dedicated transmission line has been explored in case of a few CPPs. The option of 'dedicated transmission line' needs to be tested on technical grounds as well as commercial considerations. *Prima facie* it does not appear to be the feasible alternative on account of following factors:

- Under an integrated grid operation scenario, while legally it may be feasible to develop a 'dedicated transmission' line, it would be difficult to ensure 'dedicated' nature of its operations. In interconnected transmission system, the flow of power or variations in flow in any element shall affect the power flows in other parts of transmission system. Thus, operational command and control of such transmission element will have to be undertaken in co-ordination with STU/SLDC. Thus, 'dedicated transmission line' may be feasible only in case of limited 'point to point' transmission typically, on stand alone basis.
- Further, cost of 'dedicated transmission' lines will have to be recovered only from 'dedicated users' of such transmission line which may be expensive if the 'dedicated transmission line' is very long.

Hence, it does not appear to be economical to develop long 'dedicated transmission lines' unless there is any technical compulsion or a specific commercial consideration to do so.

Wheeling of Power using Open Access

Open access mechanism provides freedom to consumers to source power from their choice of power supplier. Open access allows a bulk consumer, according to the framework developed by the appropriate Regulatory Commission, to contract directly with the generating company or with any other source of supply (other than the incumbent distribution licensee in whose area the consumer is situated). The open access framework also offers such freedom to generating company to supply power to such consumers who are eligible to avail open access. The Electricity Act, 2003, vide Section 2(47) defines open access as under:

> "Open access" means the non-discriminatory provision for the use of transmission lines or distribution system or associated facilities

> with such lines or system by any licensee or consumer or a person engaged in generation in accordance with the regulations specified by the Appropriate Commission;

Further, as per provisions of Section 9(2) of EA 2003, the captive generating plant has the right to open access for the purposes of carrying electricity from its captive generating plant to destination of its use. The relevant extract of Section 9(2) is as under:

> (2) Every person, who has constructed a captive generating plant and maintains and operates such plant, shall have the right to open access for the purposes of carrying electricity from his captive generating plant to the destination of his use:
>
> Provided that such open access shall be subject to availability of adequate transmission facility and such availability of transmission facility shall be determined by the Central Transmission Utility or the State Transmission Utility as the case may be.

Further, Section 42 (2) of the Electricity Act, 2003 stipulates that surcharge (for meeting the requirements of cross-subsidy) in addition to charges for wheeling shall not be leviable in case open access is provided to a person who has established a captive generating plant for carrying electricity to the destination of his own use.

However, for transmission of power under open access route, the various kinds of charges and losses are payable depending upon the use of transmission and/or distribution network, as the case may be. These charges are as follows:

1. Transmission charges (in case of transmission open access),
2. Transmission losses (in case of transmission open access),
3. Wheeling charges (in case of distribution open access),
4. Wheeling losses (in case of distribution open access),
5. Additional surcharge,
6. Connectivity charges,
7. Balancing market (UI) charges,
8. Operating charges (SLDC charges),
9. Default supply charges.

State Government Taxes/Duties on Captive Generation

The Tariff Policy cautions against the imposition of higher duties on captive consumption of electricity. The policy also provides for facilitating harnessing of surplus captive generation. However, various state governments have been levying duties on generation and/or consumption of power from CPPs. These duties/taxes are summarised below:

i. Andhra Pradesh: 25 paise/unit.

ii. Assam: 10 paise/unit.

iii. Gujarat: 40 paise/unit on self-consumption by CPP and rate applicable in accordance with tariff schedule, on consumption.

iv. Madhya Pradesh: Electricity duty on consumption/self-consumption @ 8 per cent of applicable tariff and cess of 10 paise/unit.

v. Chhattisgarh: Electricity duty on consumption/self-consumption @ 8 per cent of applicable tariff and cess of 10 paise/unit on generation.

vi. Orissa: Electricity duty on consumption—20 paise/unit.

vii. Tamil Nadu: Electricity duty on self-consumption—10 paise/unit.

viii. Karnataka: Tax at 5 per cent of bill amount.

ix. Uttarakhand: Electricity duty on generation—2 paise/unit.

x. Maharashtra: Electricity duty on consumption/self-consumption—30 paise/unit.

xi. Meghalaya: Electricity duty on generation/consumption/self-consumption—6 paise/unit.

Coal Linkage Policy for Captive Power

Projected Coal Requirement

As discussed earlier in the paper, coal is the major fuel source for CPPs in India and constitutes around 42 per cent of total 32,000 MW captive mainly because it is the most economical source for captive power generation. Though the share of coal based CPPs as percentage of total captive capacity has reduced marginally, they are likely to constitute a major share of the total installed captive capacity. The projected coal requirement for captive power generation is given below:

Figure 13.7

Projected Coal Requirement for Captive Power Plants

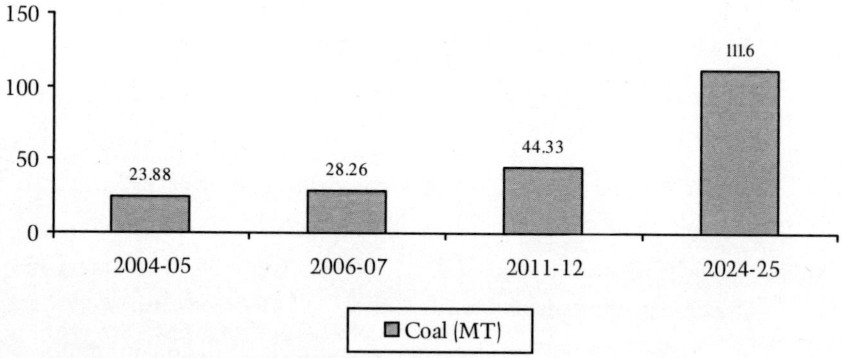

Source: India Infrastructure Report.

Coal Linkage Policy

The new coal distribution policy formulated by Ministry of Coal, Government of India in October 2007 dispensed with the previous classification of consumers into core and non-core sector and provides for treatment of consumers on the basis of merit. The coal linkage system has been replaced with a more transparent bilateral contractual enforceable agreement called Fuel Supply Agreement (FSA).

For CPPs, supply of coal has been facilitated in terms of the recent coal distribution policy as well as the policy of allocating captive coal blocks. As per the policy:

> 100 per cent of the quantity as per the normative requirement of the consumers would be considered for supply of coal, through Fuel Supply Agreement (FSA) by Coal India Limited (CIL) at fixed prices to be declared/notified by CIL. The units/power plants, which are yet to be commissioned but whose coal requirements has already been assessed and accepted by Ministry of Coal and linkage/Letter of

Assurance (LOA) approved as well as future commitments would also be covered accordingly.

The policy envisages a FSA in which both the supplier and consumer have an obligation to honour the contractual agreement.

The policy for new consumers stipulates that:

> The Letter of Assurance (LoA) to be issued now pursuant to the new policy will have a validity of 24 months for consumers/applicants of Power Utilities, CPPs & IPPs and 12 months for other consumers instead of 30 months as earlier. The allotter of LoA would be required to fulfill certain stipulated conditions and meet the milestones within this period and there upon approach coal companies for entering into FSA. Such FSA would be completed within three months. Further, with a view to ensure that only serious and committed consumers approach for LoA, they would be required to furnish an "Earnest Money Deposit (EMD)". EMD can be in the form of Bank Guarantee and would stand discharged once FSA is concluded within the stipulated period. However, on failure, the EMD will be forfeited. The amount of EMD could be initially kept at 5 per cent of the value of Annual coal requirement. However, CIL may decide a different level, based on various relevant facts, with the approval of Board of Directors of Coal India Limited.

The various milestones to be achieved within 24 months from the date of Letter of Assurance are given in Table 13.1.

Allocation of captive coal mines provides a greater scope for captive power companies in terms of the quality and quantity of coal. In recent years, there has been a rush to seek allocation of captive coal mines. Sectors such as cement, steel and aluminium are the prominent sectors which are pursuing allocation of coal mines.

Utilisation of Surplus Captive Power

As per the National Electricity Policy, a large number of captive and stand-by generating stations in India have surplus capacity that could be supplied to the grid continuously or during certain time periods. These plants offer a sizeable and potentially competitive capacity that could be harnessed for meeting demand for power. Under the statutory and regulatory provisions, the CPPs can sell up to 49 per cent of power generated by them either to the utilities or to the eligible open access consumers. The market mechanism for sale of power from CPPs has been deliberated in the separate theme paper. However, there are other ways of utilising the idle captive capacity and one such innovative successful story is CII Pune Model.

Table 13.1

Milestones in Power Sector to be Achieved by LOA Holders

Time Limits as from Date of LOA	*Target Activities as Milestones*	*Supporting Documents*
6 months	In principle approval of the Board of Directors of investment decision, in principle water allocation, Land Acquisition Notification (Section 6)	Certified copy of the decision of the board, relevant letter from appropriate state authority, relevant notification
12 months	Firm water allocation, in-principle environmental clearance, recommendation for forest clearance from state government	Relevant letter from appropriate state authority, relevant letter from state/Central government departments
18 months	Final environmental clearance, land in possession for main plant and coal handling plant	Relevant letter from state/Central government departments, possession certificate from the competent revenue authority
24 months	Forest clearance, financial closure, land in possession—90%	Letter from competent authority, letter from bank/financial authority, possession certificate from the competent revenue authority

CII Pune Model: Example of the Best Utilisation of Captive Power

Conceived in 2004 and made operational from 4 June 2006, the popularly known 'Pune Model' steered by Confederation of Indian Industry (CII) has resulted in the city of Pune being completely free from load shedding for around 1.5 years since its implementation. For many months and years prior to that, Pune was facing load shedding regularly from one to four hours a day as was the case with every other city and village in Maharashtra (with the exception of Mumbai which is not in the purview of Maharashtra State Electricity Distribution Co. Ltd. (MSEDCL)). This innovative model was based on some fundamentals which included:

- Public-private partnership (PPP) pattern where the civil society and industry work with government agencies to solve problems.
- Mitigation of load shedding in Pune not at the expense of any other community.
- The incremental cost of this initiative paid for by the beneficiaries (citizens of Pune in this case). Hence a 'pass through' for distribution utility MSEDCL with no financial burden on the state of Maharashtra.

After an elaborate public process conducted by Maharashtra Electricity Regulatory Commission (MERC) which involved, amongst others, three Public Hearings, CII's proposal was accepted by the citizens and approved by the Regulatory Commission.

How the Pune power arrangement was structured is a study in elegance. The companies having CPPs agreed to use their gensets (based on liquid fuel) during peak hours of load shedding, so that they do not draw the power from the MSEDCL grid to the extent of around 90MW which could be available to the citizens of Pune. In return for the units generated and consumed by the CPPs during the agreed hours, the CPPs were reimbursed the incremental cost, that is, the difference between the variable cost of running CPPs and the applicable MSEDCL tariff.

Based on the prevalent fuel prices, weighted average variable cost of generation as approved by the MERC worked out to Rs 10.18/kWh. Considering the average tariff for high-tension industrial users as Rs 4.04/ kWh; the reimbursement required per kWh generated and consumed by the CPP worked out to be Rs 6.14 /kWh. When this cost was spread over all consumers of the Pune urban circles which enjoyed the benefit of zero load shedding, it worked out to be approximately 40 paise per unit. This additional tariff called 'reliability charge' was levied only on those consumers which consumed 300 units per month and more. Consumers consuming less then 300 units per month were exempted from the reliability charge. Out of the targeted 6 lakh consumers in Pune`s urban circles, only 2 lakh recorded a reading of 300 units and above. Thus, only they were liable to pay the reliability charge for uninterrupted power supply. Further, the cost recovery from citizens was so framed that it was both 'progressive' as well as conducive to 'conservation'. Therefore small and marginal consumers got the benefit of '24 X7' power without having to pay a premium. This further encouraged consumers to bring down their monthly consumption below 300 units so that they don't have to pay the extra charge.

Section VI

Regulation: The Way Forward

14 Coordination in Energy Regulation

S.L. Rao

Background

The governance of the country has a built-in coordination mechanism between the Centre and the states through the National Development Council, the Finance Commission, the negotiations of state governments with the Planning Commission on the size of their plans and Central support to them, among others. Coordination between ministries in a government is through Cabinet meetings and by the Chief or Prime Ministers. Frequent meetings between ministry departments take place to achieve coordination. However the coordination has been weak and in key areas it has been time-consuming and infructuous. A good example is that of energy and within that, electricity (or power). Coordination between the Centre and the states in electricity was for long erratic and inadequate until the introduction of the Accelerated Power Development and Reform Programme (APDRP). By introducing a programme with substantial rewards and penalties in relation to an agreed memorandum with milestones for achievement it has enabled substantial progress in some states.

There are separate ministries for many different types of energy—atomic, from non-conventional sources, thermal and hydroelectric, apart from oil and gas and for coal. Power is at present the only part of the energy sector in India to have independent regulatory mechanisms (covering licensing, trading, tariffs, transmission and distribution), and separately for the Centre and the states (power is a concurrent subject in the Indian Constitution). There is a Central Electricity Regulatory Commission CERC, and there are individual state commissions (SERCs), though smaller states and union territories are permitted to have joint commissions. The

CERC regulates the Central government-owned sector and inter-state issues; the SERC is responsible for matters within a state.

The Electricity Regulatory Commissions Act of 1998 did not provide for any coordination between the Central and state commissions, a glaring omission, since the imperatives of a national interconnected grid demand it. However, the Electricity Act, 2003 (Section 66) attempts to provide for some coordination between the Central and state commissions through two institutional mechanisms. One is a coordination forum consisting of the chairperson and members of CERC, chairman CEA, and representatives of generating companies and transmission licensees engaged in inter-state transmission of electricity and the second, a forum of regulators consisting of the chairpersons of the CERC and the SERCs. (This is in addition to the voluntary Forum of Indian Regulators (FOIR). FOIR consists of all present and former regulators, presently only in electricity, that was formed in 1998 and has been functioning ever since).

Another method that provides coordination and consistency in approach is that the Act gives the Central Commission the responsibility of laying down regulations and rules on the specified matters. These are (Section 178): grid code; rates, charges and terms and conditions in respect of intervening transmission facilities; payment of transmission charges and a surcharge for providing non-discriminatory open access; reduction and elimination of surcharge and cross-subsidies; proportion of revenues from other business to be utilised for reducing transmission and wheeling charges; duties of electricity traders; standards of performance of licensee or class of licensees; details to be furnished by licensee or generating company for determining tariff in respect of generation, transmission and distribution. The Central Commission (Section 61(a)) has also the power to specify principles and methodologies for the determination of the tariff applicable to generating companies and transmission licensees.

On the other hand, there is no statutory provision for coordination beyond electricity even with regard to the major fuels used in generating thermal power in India, namely, coal and gas. Usage for power generation is at present the single most important usage of coal and gas. However, these fall under different ministries at the Centre. Power prices are regulated, with end prices to consumers being capped by the SERCs and any increase in fuel prices will not affect end consumer prices of power. The coal and gas companies concerned, with final decision by their controlling ministries, determine the prices of coal and gas (entirely state-owned in the case of coal and largely so for gas). Price coordination between the power ministry and the others is poor and largely ineffective. The Power Regulators have no opportunity to examine the validity of fuel price increases. The Power Ministry is hence reduced to plaintive pleas to the Cabinet to prevent such unilateral increases.

Environmental clearances add substantially to power project costs for generation and transmission. Delays in clearances lead to time and cost overruns. There is little or no coordination between the different sectors. Nor is the power regulator required to enforce orders of government departmental environmental regulators on matters such as pollution standards or fly ash utilisation.

Similarly railway freight is an important cost element in coal prices and has been rising, thus affecting power costs. However the power regulator has no say in such freight increases. On the other hand, railways pay the highest costs for power as compared to other consumers. Railways have had little recourse except to set up their own captive generation. Similar is the case with gas. Again the power regulator cannot scrutinise the price of gas though it is an important element in power costs and tariffs. Prices for gas are lower for the public *versus* the private sector. Government has a gas utilisation policy that sets use priorities. The Petroleum and Natural Gas Regulatory Board (PNGRB) has no role either in such price determination nor in utilisation. These are done privately by the ministry.

There is a similar issue in hydro power when the project is linked to using water for irrigation. A share of the capital and running costs must rightly be charged for water and be reflected in the tariffs of its users. However this happens to a very limited extent and substantial unpaid amounts remain in the books of the Irrigation Department. Power is charged the full costs and the power consumer pays for much of what should have been charged for water.

The present structure in the energy sector is of substantial government ownership, control and management. This is beginning to change and by 2025 there can be expected to be a large private sector in each of the different energy types. Trading in energy has begun and markets will become important in coming years. Rural electrification is an important political issue and holds opportunities for investment and challenges for coordination and regulation.

Governance in Energy Sector

Government decision-making on energy at the Central level is distributed between the Ministry of Petroleum & Natural Gas (MoP&NG), the Ministry of Coal (MoC), the Ministry of Non-Conventional Energy Sources (MNES), the Ministry of Environment and Forests (MoEF), the Department of Atomic Energy (DAE) and the Ministry of Power (MoP). Within the Ministry of Power, the Central Electricity Authority (CEA, the technical wing) works closely with individual State Electricity Boards (SEBs) and utilities in power generation, transmission, and distribution of electricity. At the state level, there exist various departments, agencies and authorities working on various sub-sectors of energy. Ironically, the Ministry of Panchayati Raj (MoPR) functioning at the village level does not have at present any role to play in rural electrification. The responsibility for comprehensive rural electrification (including quality of power and collection) is scattered between different ministries in an uncoordinated fashion.

Being a concurrent subject under the Constitution the states share powers with the Centre only on power, not on the other energy sources. However they do share powers on environmental regulation. This distribution of powers makes coordination between the different energy sources, over the country and between states difficult. The following paragraphs describe the way in which this distribution of powers is organised at the Centre.

Environmental clearance is required for all types of power projects including nuclear, hydel and thermal under the Environment (Protection) Act, 1986 (EIA notification in 1994) from MoEF, GoI. Site clearance is required for pithead thermal power stations and hydro power projects. Public hearing is required for all power projects. There is a prescribed application form and procedure to be followed for obtaining clearance. If forest land is involved in any power projects, clearance has to be obtained from MoEF under the Forest Conservation Act. In 1997 through a notification, MoEF has delegated some powers to the state governments for environmental clearance of power projects up to certain capacity. The supreme court through an empowered committee also plays a role in clearances.

Substantial extra costs have to be incurred at times to meet environmental standards. These costs are a given for the electricity regulator but there is no monitoring to ensure that the desired pollution or ecological standards that the costs were supposed to achieve were in fact achieved. (Refer Annexure A-14.1 to see the list of clearances that are required for setting up any power plant in India).[2.a 3. Weaknesses in Present Coordination]

The following chart shows the organisation structure of government in the energy sector. In addition there is the Planning Commission. It has departments to match the ministries. The close protection of their turf by each ministry makes for ineffective coordination. It is now envisaged that the Planning Commission will play this coordinating role for infrastructure.

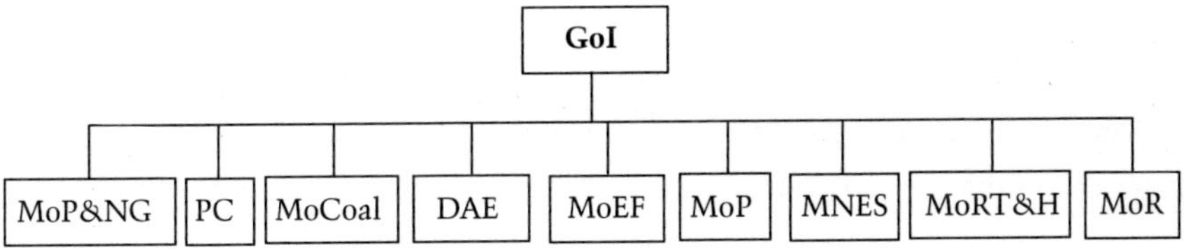

Status of Present Regulation of Coal

The Coal Ministry presently regulates coal. Nearly 70 per cent of electricity capacity in India is based on the use of coal resources and nearly 75 per cent of the coal consumed is for power generation. The sector is a monopoly of Central government-owned enterprises (Singareni Collieries are jointly owned by the Central and Andhra government). Supply, price and quality are subject to decisions by the government-owned companies with the final approval of the concerned ministry. While captive mines have been permitted for generation plants, they are very few and are not permitted to market their coal. There is said to be a draft bill for appointing a coal regulator but the draft is unavailable and we cannot say whether the regulator will regulate tariffs and quality or be responsible only for licensing captive mines and allocating supplies.

The Coal Mines (Nationalisation) Amendment Bill, 2000 (not discussed by Parliament till October 2004) explains that in view of the anticipated 'huge gap between demand and supply of coal by the end of the 10th Plan which cannot be bridged by the nationalised coal companies and the captive mining companies...it is necessary to allow the private sector to participate for non-captive purposes.' In the discussions on the Bill it has been suggested that there is a definite need to review the pricing mechanism for coal and for this purpose an independent body must regulate the price. The ministry had prepared a draft law in 1997. However it was found that the Coal Mines (Nationalisation) Act would have to be amended for the purpose. The ministry said it would re-examine the requirement and review the composition, functions and powers of the independent regulatory body. In these last seven years, no regulatory body for coal has been established.

A new policy giving captive mines to major users, subsequently the allotment of mines for Ultra Mega Power Projects (UMPPs), and permission to them to use any surplus coal for power generation for merchanting is the first dent in the principle of nationalisation. However captive mines are not permitted to sell their coal in the market. The state-owned companies in consultation with the Coal Ministry determine prices for coal from the nationalised mines. The principal user is the electricity thermal generating sector and it has no forum in which to participate in the decisions on prices, quality and even supply.

The coal industry claims that coal at the pit head is the cheapest fuel option but that it becomes very costly by the time it reaches the user at a distance because of high rail freight, royalty payments and taxes. These costs also need to be determined in an open and transparent way so that end prices of coal do not lead to sharply rising power tariffs. However, the Railways Ministry has adamantly refused to consider setting up an independent Railway Tariffs Regulatory Authority.

Status of Present Regulation of Oil and Gas

The Petroleum and Natural Gas Regulatory Board is appointed by the government. At the same time the regulator is barred from government or commercial employment for two years after he ceases to be a regulator. This seems rather onerous and might put off suitable candidates.

The Board concentrates on transportation and marketing but does not have tariff-setting powers. The regulator has powers to license the marketing of notified petroleum and gas products, establishing and operating LNG terminals, to lay, build, operate and expand as common carrier, declare designated pipelines as common carrier, regulate access and transportation rates to the common carrier, while funding is to be independent and not part of the government budget.

For notified products the regulator is to ensure adequate availability, ensure price information at retail level, prevent profiteering, secure equitable distribution and maintain an information data bank. Government can give directives to the regulator and is required to consult him before doing so but only 'where possible'.

As far as enforcement powers are concerned the Regulatory Board can use the CBI for conducting investigations, a useful provision not in the legislation for other independent regulatory bodies. It can impose civil penalty of Rs 1 crore and Rs 10 lakh per day, and for profiteering impose a penalty up to five times of unfair gains made or Rs 10 crore whichever is higher. The chairman of the Regulatory Board is to be selected by a Search Committee composed only of secretaries to government, and government is to decide on termination of the regulators. The regulator does not have powers in the present Draft oil and gas regulatory legislation to regulate trading, markets, bulk use, grid code, ownership of pipelines, determine gas tariffs, etc.

Regulation and Coordination among Fuel Transporting Ministries

The energy sector is heavily dependent on railways (for transportation of coal), sea freight (for oil and gas) and the Finance Ministries of Central and state governments for more uniform and reasonable rates of taxation.

As far as railways are concerned There is no transparency in the fixation of coal freights by rail. It would be useful for there to be an independent Railway Rates Authority, which should seek advice from the CERC and other energy regulators on the impact of rail freights on energy costs. However, the Railways Ministry is opposed to the idea that independent regulators might determine rail tariffs on the grounds that outside bodies would not appreciate the public interest and would be taking powers away from the minister who might have 'social considerations' in view.

As far as sea freight is concerned, this has become subject in recent times to severe fluctuations because of the fast growing demand in China and India (and the recent global downturn) for imported oil, gas and coal. Several actions are required to be taken by the government. These include:

- Long-term arrangements for oil and gas imports instead of leaving it to private investors in LNG terminals as in the case of gas.
- A National Shipping policy that would leave India less at the mercy of opportunistic and volatile world shipping freight rates.
- Develop risk management strategies and strategic storage capacities in the country.

Sea freight for public sector undertakings importing oil and gas in past years have been negotiated by TransChart, a no-fee non-profit unit of the Shipping Ministry. It is estimated that oil companies have saved Rs 2,000 million over three years (to 2003) on sea freight because of this centralised negotiation. TransChart has a diminished role in recent years. The electricity regulator could play a role in opening this subject since it has a direct bearing on generation costs by raising the costs of fuels.

Coordination in Taxing Energy Sector

There are many issues in the case of taxation on energy. There is no accurate estimation of the impact of taxation at different stages on the various elements in the energy sectors and this estimation has to be made so that some uniformity in taxation and impact can be introduced. The tax benefits available to generating companies under Section 10-23G has to be compulsorily passed on by the lender either in the form of lower interest rate or rebate.

The same logic holds good for minimum alternate tax (MAT). Under the two-part tariff mechanism for the computation of income tax, it is grossed up for determining the passing up of liability of the Electricity Board to the generating company to ensure that the return on equity is on net on tax basis. The applicable rate for MAT is 8.25 per cent. It gets grossed up and the impact comes to 12 per cent in the form of tariff to the Electricity Boards. This is again passed on to the customer or the government in the form of subsidies that the Electricity Boards collect from the government. 85 per cent of the money collected by the Centre is passed back to the states. So the state pays the net MAT at a very high premium to collect that 85 per cent.

Taxation on fuels is very heavy and different state governments charge different taxes. Government needs to rework the fiscal framework. Sales tax is the 'bug-bear' for LNG and the entire gas business. Tax costs are loaded into the tariff. The consumers cannot take the benefit even though it is being paid for in the energy chain. Value added tax (VAT) needs to rationalise the impact of sales tax. Reforms in taxation through the goods and services tax (GST) might minimise differentials between various states and attempt for a harmonisation of rates of sales tax between the states is required.

Naphtha and gas should be declared as 'declared goods' and therefore subject to only the Central Sales Tax (CST) rate rather than continuing with 22 per cent. Electricity must be treated as zero-rated goods so that all taxes paid up to that point could be claimed back.

Import terminals and pipelines must get equal fiscal treatment. They should be recognised as infrastructure facilities and get lower customs duty. There is need for improvement in efficiency of tax collections at all levels. A way must be found to bring in a modified VAT instead of sales tax to exempt customs duty on crude oil for power generation. Giving a specific duty exemption to crude oil for power generation is worth considering, as there is no question of leakages. A sales tax holiday for five years for new power projects might also be examined.

The total impact of taxes, in case of coal is estimated to be 23.2 paisa/kWh, incase of naphtha 67 paisa/kWh and in case of gas we get a variation of 25 paisa to 43 paisa per kWh. Exemption of customs duty should be offered to power projects as a whole and not just mega and ultra mega projects. Deemed export benefits should be extended from mega power projects to all power projects. States should be requested to exempt fuel from sales tax and electricity duty for power plants, including transmission and distribution.

Regarding tax incentives the retail investors should get tax holiday benefit made available to them as it is available to power companies. IT Section 10(23G) benefits should also be made available to retail investors. Hedging cost should be allowed for tariff (to which CERC has now agreed). Stamp duty is a major cost in new projects. Stamp duty exemption should be considered. Research and development (R&D) cess on development and design from origination and coverage (O&C) might be waived off. Exemption of MAT for projects subjected to tax holiday needs to be made. 10(23 G) is available to only those projects which are commissioned before March 2006. Hence the existing projects may not be able to take its advantage.

This exploration of tax anomalies and distortions shows the extent of the uncoordinated tax situation in the energy sector. Clearly the independent regulator cannot resolve them. But these are issues that the electricity or the gas regulator when appointed must become familiar with. The role of these regulators on these issues could be in their exercise of their prerogative of giving 'aid and advise' to the concerned government on these matters.

Concurrent Powers on Electricity

The scheme of distribution of powers between the Centre and the states in the Constitution of India (Articles 245-246) is as follows:

- The constitutional provisions in India on the subject of distribution of legislative powers between the Union and the states are spread out over several articles (Articles 245-254). However, the most important of those provisions i.e., the basic one—is contained in Articles 245-246.

  Article 245 provides, *inter alia*, that (subject to the provisions of the Constitution).

(i) Parliament may make laws for the whole or any part of the territory of India and

(ii) The legislature of a State may make laws for the whole or any part of the State.

- Thus, Article 245 sets out the limits of the legislative powers of the Union and the states from the geographical (or territorial) angle. From the point of view of the subject matter of legislation, it is Article 246 which is important. Article 246 reads as under:

246(1) Notwithstanding anything in clauses (2) and (3), Parliament has exclusive power to make laws with respect to any of the matters enumerated in List 1 of the Seventh Schedule (in this Constitution, referred to as the "Union List").

(2) Notwithstanding anything in clause (3), Parliament, and subject to clause (1), the Legislature of any State also, shall have power to make laws with respect to any of the matters enumerated in List III in the Seventh Schedule (in this Constitution, referred to as the "Concurrent List").

(3) Subject to clauses (1) and (2), the Legislature of any State has exclusive power to make laws for such State or any part thereof with respect to any of the matters enumerated in List II in the Seventh Schedule (in this Constitution, referred to as the "State List").

(4) Parliament has power to make laws with respect to any matter for any part of the territory of India not included in a State, notwithstanding that such matter is a matter enumerated in the "State List".

(5) Article 254: Repugnancy

- Since the Concurrent List—Article 246(2)—gives power to two legislatures, a conflict can arise between laws passed on the same subject by the two legislatures. To deal with this situation, article 254 of the Constitution makes the following provision:

254(1) If any provision of a law made by the Legislature of a State is repugnant to any provision of a law made by Parliament which Parliament is competent to enact, or to any provision of an existing law with respect to one of the matters enumerated in the Concurrent List, then, subject to the provisions of clause (2), the law made by Parliament, whether passed before or after the law made by the Legislature of such State, or, as the case may be, the existing law, shall prevail and the law made by the Legislature of the State shall, to the extent of the repugnancy, be void.

(2) Where a law made by the Legislature of a State with respect to one of the matters enumerated in the Concurrent List contains any provision is repugnant to the provisions of an earlier law made by Parliament or an existing law with respect to that matter, then, the law so made by the Legislature of such State shall, if it has been reserved for the consideration of the President extending beyond the State.

Finally, as envisaged by the Joint Select Committee on Indian Constitutional reforms, some entries in the Concurrent List take into account the fact, that (in future), a need may arise to enact legislation providing for mischiefs arising in the provincial sphere, which extend, (or are likely) to extend, beyond the boundaries of a single province. Examples of entries in the Concurrent List, which illustrate this consideration, are the following:

Entry 3. Preventive detention.

Entry 5. Marriage etc.

Entry 15. Vagrancy; nomadic and migratory tribes.

Entry 21. Commercial and industrial monopolies, etc.

Entry 25. Education, including Universities, etc.

(This entry was revised in 1977).

Entry 33. Trade and commerce etc., in certain products and goods.

Entry 35. Mechanically propelled vehicles etc.

Entry 38. Electricity.

Incidentally, the wisdom of including 'electricity' in the Concurrent List (Entry 38) is amply demonstrated by the successive statutory measures enacted on the subject in India after independence beginning with the Electricity (Supply) Act, 1948, supplemented or modified by a mass of recent legislation in the subject.

It is seen that on a fairly large number of occasions in the past, it became necessary to amend the Constitution, by transferring, to the

Concurrent List, certain matters from the State List. Obviously, it was thought that the matters in question were not being adequately dealt with (or could not be adequately dealt with) by the States [Entries 11A, 13A, 17B, 25 and 33A].

In conclusion, it is clear from this discussion that the Centre can legislate on electricity and the Central legislation shall prevail over that of the state. Electricity was perhaps on the Concurrent List because at the time the Constitution came into force (1951), the technology for sending electricity over long distances in a national grid did not exist. Ideally, it would be desirable to have electricity in the Central list but functioning through state level entities. Though electricity could have been taken off the Concurrent List, the Electricity Act, 2003 has avoided doing so. It has instead legislated in a roundabout way on matters within state jurisdiction. For example,

- By the flexible definition of 'captive generation',
- Rules to be laid down that would be followed by state regulatory commissions,
- The creation of a national forum of chairmen of regulatory commissions of which the chairman would be the chairman of CERC,
- Determination of surcharge for open access to be on principles laid down by CERC.

Central laws on a concurrent subject will override a state law. However, given the nature of coalition governments in which state level political parties have significant influence it is very unlikely that any Central government will exercise such powers to override the state legislation. Since the Electricity Act, 2003 gives some coordinating powers over 'Power' to the CERC, it could exercise these powers comprehensively to avoid conflicts and inconsistencies over the country.

Regulatory Principles

a. There are sectors in energy that are regulated departmentally by government—namely, renewable energy and cogeneration.
b. The independent regulatory commissions in some infrastructure sectors (telecom, electricity, tariffs of 'major' ports, controlled by the

Central government) date back to 1994. Electricity in India is in the Constitution, a subject over which both the Centre and the states have legislative power. However, Central legislation takes precedence over state legislation. We therefore have a Central electricity regulatory commission and about 24 or so state commissions, covering all the states of India. There are—Independent Regulatory Commissions: CERC; 20 or so SERCs (with a combined one for nine union territories); and the downstream Oil and Gas Regulatory Board. Atomic Energy Commission also has its own Act of Parliament.

c. Government departmental or controlled regulators: Director-General of Hydrocarbons (for upstream regulation); coal; tariffs for oil and gas and renewable energy.

Regulatory Legislation: Electricity

(See Annexure A-14.1)

i. Central Electricity Regulatory Commission (CERC)

a. Regulates central government owned undertakings (NTPC, NHPC, NLC, Power Grid Corporation, REC but not AEC), determines their tariffs;
b. Inter-state transactions—trading, transmission, exchanges and spot and futures trading;
c. Adjudicating inter-state disputes;
d. Specifies grid code.

Note: When I was the regulator I also started determining advance tariffs for a new project. This was to help developers raise finances.

ii. State Electricity Regulatory Commissions (SERCs)

a. There is one for each state and nine for the union territories. While the SERC determines the tariffs of electricity generated within the state,
b. it has to approve the tariff paid for electricity bought from outside the state. So even when CERC has approved the tariff for a Central generating company, the SERC could change the tariff that the local distributing company is allowed to pay.

c. The SERC approves purchases by the distribution company in relation to its estimated requirements for the period. The SERC will approve the purchases up to the point where the demand forecasts are met. It will do this by stacking the offers, with the least cost offers being at the bottom (merit order) and hence the first to be procured. It is possible therefore for the more expensive ones that are beyond the forecasted requirements, to fall off the stack and not be procured.

d. The SERC is also to approve open access and decide on a surcharge to be paid over the transmission tariff.

e. Appellate Tribunal for Electricity. This body will hear appeals on Orders of the ERCs. (See Annexure A-14.2)

Regulatory Legislation: Atomic Energy

(See Annexure A-14.3)

a. In the Atomic Energy Commission (AEC) 1962, amended in 1987 and the Electricity Act, 2003, as far as regulation and tariffs of atomic energy are concerned the AEC will 'coordinate' with the CEA (and later the CERC) whose decision shall be final.

Note: But that is not what the chairman of the AEC wrote to me when I was chairman of CERC and came out with a Grid Code for India, pointing out that the grid would not be able to distinguish between atomic and other power. He said that they would follow all our rules but would not 'obey' us.

b. The Electricity Act, 2003 also has similar provisions, excluding atomic power from the jurisdiction of the CERC. (See Annexure A-14.3)

Tariff Determination

i. The policy is required to reconcile two conflicting objectives: stimulate growth of the sector which means encourage investment, and safeguard the interests of the consumer.

ii. Electricity tariffs are determined under the tariff policy of the GoI. 'Accelerated growth of the generation capacity sector is essential to meet the estimated growth in demand. Adequacy of generation is also essential for efficient functioning of power markets. At the same time, it is to be ensured that new capacity addition should deliver electricity at most efficient rates to protect the interests of consumers'.

- Procurement of power:
  a. Government policy states that power procurement for future requirements should be through a transparent competitive bidding mechanism using the guidelines issued by the Central Government. These guidelines provide for procurement of electricity separately for base load requirements and for peak load requirements. This would facilitate setting up of generation capacities specifically for meeting peak.
  b. A two-part tariff structure should be adopted for all long-term contracts to facilitate merit order despatch. According to National Electricity Policy, the Availability Based Tariff (ABT) is to be introduced at state level by April 2006. This framework would be extended to generating stations (including grid connected captive plants of capacities as determined by the SERC). The Appropriate Commission may also introduce differential rates of fixed charges for peak and off-peak hours for better management of load.

Power Purchase Agreements

(i) Should ensure adequate and bankable payment security arrangements to the generating companies in case of persisting default in spite of the available payment security mechanisms. Accelerated growth of the generation capacity sector is essential to meet the estimated growth in demand. Adequacy of generation is also essential for efficient functioning of power markets. At the same time, it is to be ensured that new capacity addition should deliver electricity at most efficient rates to protect the interests of consumers. This policy stipulates the following for meeting these objectives.

ii. In practice this has led to the evolution of different tariff methods available to generating companies:

a. Power purchase agreements (PPAs) with inter-state selling price decided by CERC and state level purchase price by distributing company decided by SERC so that the tariff is in the merit order.
b. Tariff based competitive bidding: This is usually with special fiscal preferences and in the case of domestic coal, with the allocation of captive coal mines from those available with the nationalised coal companies.
c. Merchant Power-Spot, medium-term (month) and soon futures contracts; also long-term purchase agreements. The idea here is that these tariffs are outside the regulators' purview. Private long-term purchases are possible. However the tariffs on such contracts will be much lower than available on opportunistic spot or medium-term transactions.
d. Unscheduled interchange (UI) surcharge is a penal charge enabling a commercial mechanism to stabilise frequency in the grid. It is also used as price by some unscrupulous state agencies that starve their customers to take advantage of these high penal charges and helping to stabilise the frequency.

iii. Annual revenue requirement (ARR): As far as retail tariffs are concerned, they are determined by the SERCs on the basis of an ARR. Various norms are in force for operating norms; capital costs; RoE; fuel escalation costs. Today such information on atomic energy is not in the public domain.

Subsidies

Subsidies are specifically mentioned in Section 65. The provision relates to reimbursement of subsidies to the distributing companies who are required to sell below costs to certain consumers. There is no provision regarding Central government reimbursing subsidies. If nuclear power is more expensive, there may be need for such reimbursement by Central government to enable it to be on the merit order for SERCs. (See Annexure A-14.4)

Nuclear Electricity Regulation

When I was a Central regulator and issued the first national Grid Code, I asked the chairman of the AEC as to how he proposed to operate in the circumstances of the exemption to AEC from my commissions' jurisdiction. He replied that they would follow all the rules but not be subservient to my commissions. It needs to be clarified as to how this matter will be settled since there will be the existing government-owned plants in addition to any new privately-owned plants.

There is also the question of capital costs (said to be escalating @ 15 per cent per annum) and the high operating costs, said in Europe to be many times thermal generation costs in India. There is also the question of accident liability and the inclusion of that in the tariff calculations. If the end result is that nuclear energy is much more expensive than others, there will have to be a subsidy mechanism from government funds if the nuclear energy is not to fall off the electricity regulator's merit order.

Finally of course, any public-private partnership (PPP) structures must fit in to the regulatory regime of the Nuclear Suppliers Group, Atomic Energy Regulatory Board of India, International Atomic Energy Agency, as well as the Indian Electricity Act, 2003, the Indian Contract Act, 1860 and the Atomic Energy Act, 1962. If there are additional costs in conforming to these, they will also have to come into the calculations.

Desirable Changes

a. Break Ministerial Boundaries

We can do this by creating such agencies without reference to the closely held turf boundaries of ministries and departments of governments. In this way we can also deal with another fault in our governance that decisions are taken without coordinating with other aspects that might fall in the purview of other ministries or departments.

b. Non-Proliferatory and Holistic Regulation

To avoid regulatory proliferation and ensure holistic decision-making we could have a Central energy regulator instead of separate ones for electricity, upstream oil, downstream oil, gas and coal. This regulator would be responsible for tariffs and licensing of production, transmission, distribution, and supply of coal, gas and electricity. This will ensure that the interests of all can be coordinated, especially since electricity, a major user of the others, has its prices capped. Oil and oil products could be regulated separately since issues of national security and highly technical issues in oil exploration have to be considered. Retail distribution of oil and oil products has very different issues for consideration.

At the state level the state electricity commissions could also be made responsible for a closely related area, namely ground water (for drinking and irrigation) including permission for ground water exploitation, pricing, etc. Since subsidised or free electricity is responsible for the massive over-exploitation of ground water, this could help to relate the one to the other. Since sanitation is closely related to drinking water quality, the two regulators will need to meet from time to time on related issues.

The essence is to promote transparency, public participation and open decision-making wherever the public interest is involved. If government could be so in its normal functioning, such new agencies may be redundant. But governments in India have invariably been secretive, tend to engage in superficial and limited consultation and rarely give reasons for their decisions. They are unlikely to change soon.

If we do not look at such holistic responsibility for regulatory agencies we would be creating a multitude of regulatory agencies in addition to government departments, without any improvement in the quality and cost of the services they have to regulate. A regulatory morass is the inevitable consequence of regulatory proliferation. It does not have to be so.

c. Coordination

The Electricity Act, 2003 is a bridge between the present and 2025. It also gives the central commission the responsibility of laying down regulations and rules on specified matters and also to specify principles and methodologies for the determination of the tariff applicable to generating companies and transmission licensees. There is no legal provision for coordination beyond electricity even in relation to the major fuels used in generating thermal power in India, *viz.*, coal and gas. Amendments are necessary to the existing Bill to create regulatory authorities for oil and gas and to amend the Coal Mines (Nationalisation) Act. Similarly, in shipping freight rates the regulator could play a useful role in advising the government.

A mechanism will have to be legislated under which regulators for related areas meet regularly on boundary issues. This must apply to energy, water, rail, road and sea transportation, environmental regulation and Telecommunications (since power transmission wires are used to hang the telecom wires). If coal and gas are not with a single energy regulator, they must also be in this list.

d. Taxation

There are many variations between states and Central taxation is also an important component in energy tariffs. Here again some coordination perhaps as an advice to government from the regulator might be useful.

e. Subsidies and Role of Regulator

The electricity regulator must play a more proactive role than he has done in handling subsidies. He must have reliable data and take steps to get them. Thus he must estimate:

- The real amount of subsidy to agriculture.
- How much electricity supplied to agriculture contributes to GDP so that some idea can be had of the social benefit of such supplies.
- Introduce innovations in metering—11 KV metering, prepaid meters.
- Costs of supply to agriculture in off-peak hours.
- Power and water use efficiency under different tariff regimes—flat rate, metered per unit.

Clearly the regulation of agriculture tariffs will require coordination with energy efficiency and many studies and experiments.

f. Role of Regulator in Future Viability of Electricity Sector

- Regulators must focus on improving the infrastructure of connectivity, metering and equipment quality.

- They must also make serious effort to identify the population to be subsidised and develop methods to do so within the available subsidy (plus cross-subsidy when it exists).
- Energy regulators are enjoined to promote competition. The regulator must clearly define what is meant by competition, the extent to which it is possible or can be simulated and the stages through which the sector will pass before competition is largely established.
- The need for coal to be regulated independently must be questioned and it requires to be regulated together with electricity.
- Either by legislation or regulation production/generation, transmission/pipelines and distribution might be separated so that the different operations might be confined to separate corporate entities. This is important when the same entity is in more than one operation.
- The ERCs will have to work closely with the newly created regulator for energy conservation, the Bureau for Energy Efficiency.
- With more private generation capacity set to enter the grid and privatising of distribution, the ERCs must impart independence to the transmission operator and make load despatch into a neutral function.
- It might be better if operators do their own forecasts and the regulator works to a broad plan of transmission requirements than to a national estimate that is made by the CEA. Data and forecasting are areas in which electricity regulatory commissions must play a key role.
- The routing and placement of gas pipelines and terminals for LNG need more scrutiny. We need a balance between the centralised planning in electricity and the relative laissezfaire in the case of oil and gas.
- Open access must be ensured in electricity and gas and the requirement in the Electricity Act to use surcharges for meeting the cost of cross-subsidies should not be allowed to restrict open access and eliminated as soon as possible.

Changes Required in Present Independent Regulation of Energy

Present Challenge

i) Indian governance over the last 57 years has not enabled sufficient involvement and participation by the concerned people. Procedures are complex and secretive so that even many government servants find them hard to follow. Governments are unable to implement programmes faithfully, fully and without waste and corruption. There is no holistic decision-making. It is usually only sectoral. Major decisions are announced without giving reasoned arguments to show why one decision was made as against other options.

ii) There is a growing presence of the private sector in oil and gas and to a much lesser extent in power. The use of commercial energy is still low. Many households are unconnected. Many cannot afford market rates of power, kerosene and liquefied petroleum gas. Government supplies, some of them at low rates and sometimes free. Inefficiencies abound in production/generation, transportation/transmission, distribution and supply. The state governments that are the principal owners of power generation and distribution are hard pressed to find resources for other governmental responsibilities since they are spending so much of state resources on supplying cheaper power.

iii) In the circumstances we must think of ways to introduce participation and consultation into the governmental process especially in areas of infrastructure while recognising the role of government. Since there are practically no people with the required multi-disciplinary expertise, it is a thought for consideration that a regulatory cadre might be created that could staff memberships of all independent regulatory bodies. The people might be found from amongst those who have spent over 15 years in senior decision-making positions in the defined areas of expertise, given some training and then sent to function in one or the other independent regulatory commission.

Effectiveness of ERCs

i) In the present environment in the power sector the task of the independent regulator is to improve efficiencies and quality. Independent regulation exists only in power and has on the whole been able to a limited extent rationalise distortions between user groups in tariffs; at the Central level and in some states especially in cities the regulator has been able to improve quality. There is some modest improvement in T&D losses and significant improvement in plant load factors in generation. Metering has expanded though it is still only for a portion of users. However there is little change in using environmentally benign energy, targeting subsidies and keeping them in check, in moving towards cost of service for user charges, competition at least in serving bulk users, removing major transmission constraints, significantly reducing voltage fluctuations, improving safety, cutting technical losses in T&D, improving collections, imbibing professional management, commercial and entrepreneurial attitudes and mindsets. The oil and gas sectors are also beset by problems caused by subsidies and cross-subsidies and the lack of adequate pipeline capacity. Coal remains a government monopoly and tariffs and quality pose major costs for users. The task in the coming years is to correct these conditions.

Transition Directions (2005-2010)

It will take at least a decade before the electricity sector in India has a large (around 30 per cent) private sector that can push for regulatory reform as it is doing in the telecom sector. What should be the regulatory approach during the period of transition when the state-owned sector is dominating?

i) It seems apparent from the experience of the ERCs particularly the SERCs that a loose federal approach will not work. It does not fit the requirements of electricity that follows the law of physics and not of political boundaries. The power system requires up to the minute accounting, full metering, adequate T&D, good equipment maintenance, fail-safe systems to island load centres and cut load or generation as required by the system, strong grid discipline, close monitoring, efficient operations, and highly professional management. It needs cohesiveness and consistency in policies. All this demands strong centralised control and direction.

The ERCs function in a power system that is primarily state-owned. With enterprises so close to the bureaucracy and the bureaucracy's tacit support for most of the challenge and defiance that they display against orders of ERCs, independent regulators face difficulty. State ownership and dominance is unlikely to reduce in the near future. One way is for the system to be modified while retaining transparency and consultation while making the regulators part of the government. However the regulators might be given fixed tenures during which they cannot be moved. Their independence from the government can also be written into the law so that their confidential annual reports, etc., are dealt with in such a way that there is no possibility that they could be used to make the regulators more amenable. Staff could be taken from outside the government as well with less restrictive compensation policies rather than applying to normal government servants. The model of chief election officers in the states is a possible direction for development of this institution at least till such time that state ownership is dominant.

ii) It is apparent that we have too many SERCs in relation to available qualified members and staff. For they issue orders with contradictory principles on the same issues. Ideally, we should have a Central regulator with benches in each state.

Alternatively we should have regional regulators with the opportunity for each state to issue policy directives to them for their state on specified matters. This may require changes in the Constitution of India. Alternatively the precedent of the ERC Act, 1998 which was amended to permit joint commissions for small states might be extended with the consent of the states to achieve this object of reducing the number of ERCs. This will also save considerable expense. More important, such bodies can build expertise from the limited numbers of experts who are available.

iii) A stronger role for the CERC in relation to the SERCs is highly desirable since the CERC has been better able to build and use expertise. The new Act gives the CERC a definite role in relation to the SERCs unlike the ERC Act, 1998, which gave it no role or relationship at all with the SERCs. The CERC must use its powers more vigorously while also using other mechanisms like the statutory forums and the informal one of FOIR to achieve consensus with SERCs.

iv) There must also be a closer consultation between the ERCs and the concerned governments so that the ERCs march in step with the political and administrative commitments. This is also provided for in the new Act. However there must be no compromise on the commissions' independence and transparency of functioning. Perhaps this could be achieved by having open meetings between governments and commissions or at the least, ensuring that the minutes of the meetings of this new proposed forum are publicised immediately after the meetings take place.

v) Another possibility is the drafting of more detailed legislation so that the SERC functions within rigid limits set out by government. This was done in Delhi with extensive use of the state's power to issue policy directives and is proposed as part of the Karnataka reforms. In this way the SERCs would function within the policies of government.

vi) The idea of independent regulatory commissions like the ERCs is a good step forward in our system of governance. They make government more transparent, being the citizenry closer to the decision-making process and are in line with the process of decentralisation and freedom of information to enable reducing the alienation of people from the government. We should not abandon them but make them effective.

Annexure A-14.1

Provisions of Section 73, the Electricity Act, 2003:

- i) to regulate the tariff of inter-state generating companies and of central government owned generating companies.
- ii) to regulate the inter-State transmission of electricity.
- iii) to determine tariff for inter-State transmission of electricity.
- iv) to issue licenses for inter-state electricity transmission and trading.
- v) to adjudicate upon inter-State disputes.

e) to specify Grid Code.

- vi) to specify and enforce the standards with respect to quality, continuity and reliability of service.
- vii) to fix the trading margin.
- viii) it also has advisory functions, such as formulation of National Electricity Policy and Tariff Policy; promotion of competition, efficiency and economy in the activities of the electricity industry; promotion of investment in electricity industry; any other matter referred to the Central Commission by the Central Government)

13 State Electricity Regulatory Commissions—SERCs:

Provisions of Section 86, the Electricity Act, 2003:

- i) determine the tariff for generation, supply, transmission and wheeling of electricity, wholesale, bulk or retail within the State.
- ii) regulate electricity purchase and procurement process of distribution licensees.
- iii) facilitate intra-State transmission and wheeling of electricity.
- iv) issue licenses for intra-State transmission, distribution and trading.
- v) promote cogeneration and generation of electricity from renewable sources of energy.
- vi) adjudicate upon the intra-State disputes.
- vii) specify or enforce standards with respect to quality, continuity and reliability of service by licensees.
- viii) fix the trading margin in the intra-State trading of electricity.

ix) They also have advisory functions, that include promotion of competition, efficiency and economy in activities of the electricity industry; promotion of investment in electricity industry; reorganisation and restructuring of electricity industry in the State; matters concerning generation, transmission, distribution and trading of electricity or any other matter referred to the State Commission by that Government.

Annexure A-14.2

Section 120 of Electricity Act, 2003: Procedure and Powers of Appellate Tribunal

The Appellate Tribunal shall not be bound by the procedure laid down by the Code of Civil Procedure, 1908, but shall be guided by the principles of natural justice and, subject to the other provisions of this Act, the Appellate Tribunal shall have powers to regulate its own procedure. (No. 5 of 1908).

1. The Appellate Tribunal shall have, for the purposes of (No. 5 of 1908) discharging its functions under this Act, the same powers as are vested in a civil court under the Code of Civil Procedure, 1908, while trying a suit, in respect of the following matters, namely:-
   a. summoning and enforcing the attendance of any person and examining him on oath;
   b. requiring the discovery and production of documents;
   c. receiving evidence on affidavits;
   d. subject to the provisions of sections 123 and 124 of the Indian Evidence Act, 1872, requisitioning any public record or document or 1 of 1872 copy of such record or document from any office;
   e. issuing commissions for the examination of witnesses or documents;
   f. reviewing its decisions;
   g. dismissing a representation of default or deciding it *ex-parte*;
   h. setting aside any order of dismissal or any representation for default or any order passed by it *ex-parte*;

i. any other matter which may be prescribed by the Central Government. (No. 5 of 1908; No. 1 of 1872, No. 45 of 1860).

2. An order made by the Appellate Tribunal under this Act shall be executable by the Appellate Tribunal as a decree of civil court and, for this purpose, the Appellate Tribunal shall have all the powers of a civil court.

3. Notwithstanding anything contained in sub-section (3), the Appellate Tribunal may transmit any order made by it to a civil court having local jurisdiction and such civil court shall execute the order as if it were a decree made by that court. (No. 45 of 1860).

All proceedings before the Appellate Tribunal shall be deemed to be judicial proceedings within the meaning of sections 193 and 228 of the Indian Penal Code of 1974. Penal Code and the Appellate Tribunal shall be deemed to be a civil court for the purposes of sections 345 and 346 of the Code of Criminal Procedure, 1973 .

Annexure A-14.3

Provisions of Section 22, AEC Act, 1962, amended in 1987: Notwithstanding anything contained in the Electricity (Supply) Act, 1948, (superseded by the Electricity Act, 2003) the Central Government shall have authority:

(a) To develop a sound and adequate national policy in regard to atomic power, to co-ordinate such policy with the Central Electricity Authority and the State Electricity Boards constituted under sections 3 and 5 respectively of that Act and other similar statutory corporations concerned with the control and utilisation of other power resources, to implement schemes for the generation of electricity in pursuance of such policy and to operate either++ by itself or through any authority or corporation established by it or a Government Company, atomic power stations in the manner determined by it in consultation with the Boards or Corporations concerned, with whom it shall enter into agreement regarding the supply of electricity so produced;

++ Inserted vide the Atomic Energy (Amendment) Act, 1987 (No. 29 of 1987).

(b) To fix rates for and regulate the supply of electricity from atomic power stations either+++ by itself or through any authority or corporation established by it or a Government Company in consultation with the Central Electricity Authority.

(c) To enter into arrangements with the Electricity Board of the State in which an atomic power station is situated either by itself or through any authority or corporation established by it or a Government Company, for the transmission of electricity to any other State;

Provided that in case there is a difference of opinion between the Central Government or such authority or corporation or Government Company as the case may be, and any State Electricity Board in regard to the construction of necessary transmission lines, the matter shall be referred to the Central Electricity Authority whose decision shall be binding on the parties concerned.

(2) No provision of the Indian Electricity Act, 1910, (superseded by Electricity Act, 2003) or any rule made there under or of any instrument having effect by virtue of such law or rule shall have any effect so far as it is inconsistent with any of the provisions of this (AEC) Act.

(3) Save as otherwise provided in this Act, the provisions of this Act shall be in addition to, and not in derogation of, the Indian Electricity Act, 1910 and the Electricity (Supply) Act, 1948.

Provisions of Section 184 of Electricity Act, 2003

The provisions of this Act shall not apply to the Ministry or Department of the Central Government dealing with Defence, Atomic Energy or such other similar Ministries or Departments or undertakings or Boards or institutions under the control of such Ministries or Departments as may be notified by the Central Government.

+++ Substituted vide the Atomic Energy (Amendment) Act, 1987 (No. 29 of 1987).

Annexure A-14.4

Provision of Section 65 of Electricity Act, 2003

If the State Government requires the grant of any subsidy to any consumer or class of consumers in the tariff determined by the State Commission under section 62, the State Government shall, notwithstanding any direction which may be given under section 108, pay, within in advance in the manner as may be specified, by the State Commission the amount to compensate the person affected by the grant of subsidy in the manner the State Commission may direct, as a condition for the license or any other person concerned to implement the subsidy provided for by the State Government:

> Provided that no such direction of the State Government shall be operative if the payment is not made in accordance with the provisions contained in this section and the tariff fixed by State Commission shall be applicable from the date of issue of orders (9399).

15 The Way Forward in Regulation and Policymaking

Ajay Shankar

A developed India by 2020, or even earlier, is not a dream. It need not be a mere vision in the minds of many Indians. It is a vision we can take up and succeed.

— A.P.J. Abdul Kalam

A vision for India in 2020 has to be one of a developed India experiencing double digit growth rates over the decade, with the divide between Bharat and India being bridged through inclusive growth and significant reduction in poverty. The National Electricity Policy recognises that 'electricity is one of the key drivers for rapid economic growth and poverty alleviation'. An India with all households being electrified and power cuts being a matter of history is a vision which should become a reality soon and, certainly, well before 2020. With the Electricity Act, 2003, a heavy responsibility for the healthy growth and development of the sector has been entrusted to the Electricity Regulatory Commissions (ERCs) who must see it as their historic mission to transform this aspiration into a reality.

If one looks at the experience of regulation till now it can aptly be described as a case of the glass being half full or half empty depending on one's perspective. From the half full perspective the achievements over the last decade have been quite remarkable and impressive. In this context it is necessary to recall that the Electricity Regulatory Commissions Act, 1998 could go through Parliament only as an enabling legislation with the states having the option to set up a Regulatory Commission only if they wanted to. The Power Finance Corporation would offer a lower interest rate if the states agreed to set up a Regulatory Commission. When the Power Ministry began negotiating MoUs on reforms with the state governments, the setting up of a Regulatory Commission was a major negotiating issue;

the argument on behalf of many states would be that they were efficient or becoming efficient and this is what really mattered and should matter and how was a Regulatory Commission relevant or material for this objective. When the Electricity Act was being conceived and was going through the process of stakeholder consultation, moving from an optional provision to a mandatory provision, for the setting up of Regulatory Commissions, it was seen as a radical and a bold measure because only a few years back the attempt had not succeeded. All this now appears so remote. The Electricity Act, 2003 was enacted by the NDA with the support of the Congress in Parliament. The ERCs have not been only been there for some years now but are also accepted by all as being there to stay and as an intrinsic and critical part of the power sector.

Initially there were considerable concerns regarding the acceptance of the authority of the commissions as well as their autonomy and independence. Fortunately these concerns also appear dated now. The authority and independence of the commissions is now seen as a fact of life. There were related second order concerns about adequacy of resources, financial and administrative powers and possible dilution of independence and stature on account of these. By and large these issues have also been settled satisfactorily. The commissions are sufficiently empowered and, what is more important, feel so and are also perceived as such. The regulated licensees have also come to terms with the reality of the existence of the commissions and their regulatory oversight. The issue of transparency in the regulatory process has also been settled. The process of challenge of decisions of the commissions in the Appellate Tribunal by aggrieved parties has also evolved. The Regulatory Commissions and the Tribunal have set healthy standards of transparency and also speed of decision-making. That state governments would actually pay subsidies to the utility as mandated by the orders of the Regulatory Commissions had appeared a very ambitious provision. This too is now seen as being natural and normal. Grudgingly, state governments have accepted that subsidies have to be given from the budget if farmers are to be given subsidised, or, as is often the case now, free electricity. The relationship between the Central Commission and the state commissions has not been an issue. The unusual mechanism of the Forum of Regulators (FOR) is proving useful for evolving a shared understanding of the key issues and problems of the power sector in India and possible approaches towards addressing them through the regulatory process.

In light of the above discussion, then where is the glass half empty and what are the key challenges for this decade? The foremost challenge lies in the further evolution of the mindset of the regulators and acceptance of their full responsibility. At one level the regulator can see his role as a reactive one which is the normal inclination as well as the historic experience, especially of the US, where a utility comes with a tariff proposal and through a transparent quasi-judicial process the regulator takes a decision on tariff proposals. This dimension of the regulatory process is now well established in India. But the needs and expectations from the regulatory process in India are far greater. The Electricity Act mandates that the commissions would be guided by the Electricity Policy and the Tariff Policy. These envisage an India where all households have electricity and power cuts are a matter of the past. By implication the regulators have responsibility and, perhaps, the most critical responsibility, in ensuring that the vision of a transformed power sector envisaged in these ambitious policies becomes a reality. This, in turn, requires assuming at a conceptual level the responsibility for the development of the sector and, therefore, taking the initiative and being proactive rather than reactive. It means ensuring that the required investments take place. Investments need the right price signals and the generation of confidence in the markets. At the end of the day investments need to be financed, debt needs servicing and equity must appreciate and give returns by way of dividend. It also means going beyond precedents of the past and attempting things which are new, have never been done before, are difficult, innovative and break new ground.

This is very difficult in a complex sector like the power sector and that too with the requirements of stakeholder consultation and the generation of a reasonable measure of consensus. An illustrative example is the progress in implementing the concept of a framework for an all-India transmission tariff which is sensitive to distance, direction and quantum of power flow. According to the Tariff Policy such a transmission tariff framework was to be implemented by April 2006. The CERC has done commendable work in evolving this framework. But this has taken more time and effort than was initially estimated. It is good that this is now under way as this is the basic need for an efficient all-India power market where investment and procurement decisions flow from the right price signals with respect to transmission costs. Generation projects should now get located on the basis of economic efficiency. Transmission investment decisions should also become easier.

Both the CERC and the SERCs have done pioneering work in promoting renewable energy development in the country. When the provisions of the Electricity Act were being crystallised there was a feeling in some quarters that giving the Regulatory Commissions the responsibility for mandating the share of renewable energy to be procured by distribution licensees and to progressively increase this mandate was too feeble a provision for the real promotion of renewable energy. This was something for governments to do. Regulatory Commissions are expected to push costs down and moderate, if not avoid, tariff increases. Renewable energy is by definition more expensive. How could Regulatory Commissions then be reasonably expected to promote renewable energy? This was for the government to mandate by law and/or through subsidies. Now the rest of the world is taking note of the rapid growth of renewable energy in India and the wonderful work being done by our Regulatory Commissions.

The real challenges of the power sector in India lies in the area of distribution and this is where the State Regulatory Commissions face their most difficult challenges. Distribution is the retail end of the business where customers pay for final electricity consumption. This has to be healthy and commercially viable. Consumer interests need to be safeguarded. Though the SERCs have full authority for setting tariffs they operate in a social and a political environment where perceptions of inefficiency, wastage in the distribution utilities make it difficult to take decisions regarding increase in rates. However, in the absence of the necessary increases it would be difficult to sustain and increase the flow of private investment into the sector on the scale that is required. There is a growing feeling among analysts that the commissions have been too conservative in giving the tariff hikes that are needed.

The Tariff Policy had envisaged a way forward for the SERCs to overcome this dilemma through the segregation of the tariff for the wires business, i.e., distribution, from the cost of power. Open access which is a key new feature of the Electricity Act needs this as a precondition to become operational. The Tariff Policy also envisages a multi-year tariff framework. The SERCs need to move onto the next phase and should provide for a multi-year distribution tariff trajectory for only the carriage cost or the wires business. A declining trajectory, even if it be a modest one, may emerge as the appropriate initial multi-year tariff on the basis of a realistic ATC (aggregate technical and commercial) loss reduction programme. If it does, the SERCs would get public acclaim for lowering tariffs and pushing the distribution licensees to achieve efficiency gains.

With the segregation of the tariff for the wires business, the distribution licensee should be allowed to pass through power purchase costs with retrospective regulatory oversight rather than prior regulatory approval. By allowing the utilities to pass through power purchase costs, the SERCs would spare themselves the burden of taking decisions on generation/power purchase costs as a part of the tariff exercise for the wires business.

Rise in fuel costs are in any case a pass-through in generation tariffs. If fixed costs are higher in newer plants these costs are accepted in case of generation sources which are selected through tariff based bids, or, get determined through the regulatory process by the appropriate Regulatory Commission. As generation is separately subject to regulatory oversight, power purchase costs should logically be allowed as a pass-through. To ensure that the power purchase costs that are passed through to the consumer by the distribution licensee are reasonable, the requirement of *ex post facto* scrutiny by the Regulatory Commission should suffice. In any case the power purchase programme of the distribution licensee should be subject to regulatory oversight. But if this process is delinked from the tariff determination exercise, it would be easier for the Regulatory Commissions to take a fairer long-term view and for the utilities to also do so.

One implication of adopting this approach would be that it would require treating the politically sensitive agricultural and lifeline tariffs for domestic households on a separate footing with actual tariff setting by the Regulatory Commissions continuing as at present. But if the rest, which is a growing share of the total load, is enabled to have tariffs that reflect actual costs then the system moves quickly towards financial health and transparent, manageable and affordable subsidies.

Once agricultural and lifeline tariffs are treated on a different footing and the rest of the consumers are paying market prices for electricity and efficient regulated tariffs for the wires business then it would become possible to get the right price signals to get the requisite investments for doing away with energy and peaking shortages for consumers. It has been seen that investor response to credible price signals in India is now quite swift. Almost half of new generation capacity is now coming up in the private sector. Once urban consumers are paying for actual power procurement costs then it would be possible for the distribution licensee to procure expensive peaking power and give his paying customers the option of being free of power cuts and paying for it. As a first step, time of day tariffs for consumers as well as generators would need to be introduced as envisaged in the Tariff Policy. This would send the right price signals to both consumers as well as generators. Consumers could try and increase consumption during off-peak hours and lower their costs. Peaking power procurement efforts by distribution licensees would attract investments for peaking power supply capacities, be they open cycle gas-fired stations, pump storage hydro power plants, or even solar plants with storage. Even coal-fired stations can meet peak demand as they have been doing in South Africa which had no hydro power or gas. All that happens is that the PLF for peaking stations is naturally much lower and there could even be units which shut down every day during off-peak hours and are again lit for meeting peak load.

For the consumer, competitive procurement and supply of peaking power by his distribution licensee is bound to be a cheaper and a better option than having to invest in personal captive generation units which is what he is being forced to do today. Today, a customer willing to pay market prices for peaking power cannot get his distribution licensee to buy that power and supply it to him as the regulatory process does not permit this. Affluent neighbourhoods, industrial and commercial areas are all likely to be willing to pay what it costs to be free from power cuts because they are in any case paying more today through captive generation. It is for the Regulatory Commissions to not come in the way and to permit market forces to work efficiently so that supply meets actual demand. With a robust smart grid, commercial and affluent residential areas could choose to pay more and be free of power cuts. Neighbourhoods of weaker sections could opt out of expensive peaking power and live with some power cuts. Industrial establishments could choose to work only during off-peak hours if the price differential between peak and off-peak power is large enough. Load management by consumers would lower peak demand and make provision for adequate supply easier. The key is to introduce market based price signals.

India could well be free of power cuts for paying consumers in about six years from the date that the Regulatory Commissions put in place the right regulatory framework for this purpose. This is not an unrealistic assessment if one factors in recent experience of project execution time and pre-project preparation including financing time.

With separate tariffs for the wires business alone, investments in making the grids smarter and more robust would now make financial and commercial sense and could be expected to take place. The importance of investment in the distribution infrastructure is not adequately appreciated or recognised because of the persisting shortages in generation. The fact of the matter is that in most cities the distribution infrastructure cannot handle peak summer loads. The system is usually not to put to test as there is not enough peaking power for meeting peak loads but if it were available, the network would not be able to handle it. Separation of agricultural pump set feeders would also make commercial sense. Governments could then ration power supply in a micro watershed basin to ensure sustainable ground water usage. Ensuring ground water availability for future generations is the real challenge. In fact many would argue that this is the most important and immediate environmental challenge confronting India. Finding the money in the budget for the power subsidy is actually a second order issue in comparison. Efficient rationing of power supply for irrigation would in addition to maintaining ground water also reduce the subsidy bill for the government. Time of day generation tariff would also lower the agricultural subsidy bill for the government. The power sector could also become a pioneer in moving subsidies from the utility to the actual consumer through smart cards. The farmer given the right incentives through the smart card and a smart grid may actually consume less electricity and, what is even more important, less water than he is doing today. For this the Regulatory Commissions would need to take the initiative and provide the leadership. The state governments would welcome such leadership. Ultimately the farmers would be better off. As this would not be about raising tariffs for agriculture but about bringing efficiency and reliability of supply to ensure timely and adequate availability of water when needed for crops and also sustainability in ground water usage, this initiative should not generate resistance and, may, even, be welcomed by farmers.

One of the difficulties that the Regulatory Commissions have been experiencing is the somewhat opaque nature of the data which the distribution licensees bring before the commissions. This is partly due to the problems in generating relevant data internally. In fact this is a perennial problem for senior management in the utilities. The Power Ministry's reform MoUs had the stipulation of energy audit as a standard feature. The APDRP programme has had this as an intrinsic requirement. But in the absence of a grid that is smart enough and adequate, regulatory oversight progress, so far, has not yet been fulfilling. Once the Regulatory Commissions have got distribution tariffs right and have got the wires business on a sound commercial footing, then they can insist on the necessary IT investments to make the grids smart enough. Smart grids lend themselves to being made more robust at lower costs. This would also be the key to making open access technically a feasible and workable proposition. Ultimately open access has to take place through the wires of the distribution licensee and this needs a robust and a smart enough grid. How important robustness is can be gauged by the recent experience of Washington DC in this year's severe winter snow storm where over 400,000 thousand consumers went without electricity. Restoration took days as the grid was evidently not robust enough and requisite investments had not been undertaken. Availability of open access is of no value if the local network has broken down. This happens often enough in Delhi on really hot summer days as the air conditioning load reaches higher peaks every year.

The Regulatory Commissions also need to put in place arrangements for an independent technical audit to be done periodically. Such an audit is envisaged in the Tariff Policy and is the key to more effective regulation. Independent financial audit by chartered accountants is considered a normal way of getting a sense of the real state of accounts. Similarly, independent technical audit is essential for the Regulatory Commissions to get a sense of the real state of the distribution network. This would also make easier the investments for modernisation. Mandating a time path for actual enforcement of reliability standards with penalties would also then become feasible. The benefits of competition through open access would also become a real rather than a theoretical possibility.

With greater transparency through independent technical and financial audit, the Regulatory Commissions would be better placed to play a leadership role in bringing greater efficiency and rationality into power subsidies. Highly subsidised power supply for agriculture and for lifeline domestic consumption would remain a political necessity and therefore a reality for the foreseeable future in India. In retrospect it can be said that time and effort was needlessly spent in not accepting this and trying to get agricultural tariffs to move up. With smart grids, energy audit, time of day generation and distribution tariffs, it should be possible to get real costs and hence shadow prices for the supply of electricity to a domestic lifeline consumer in a small village habitation and an agricultural pump set. The economics of decentralised generation through solar and biomass would appear quite different if shadow prices actually reflected real costs. The economics of local storage of surplus off-peak power to be able to cater to evening demand would also appear in an altogether different light then. Storage technology is experiencing considerable R&D efforts globally thanks to the electric vehicle. This has *prima facie* relevance for small loads in rural settlements. Actual evening supply for lighting and TV instead of power cuts which is the real experience of so many electrified villages in India would transform the quality of their lives. School children would actually be able to read in the evening and at night with electric lights.

The Electricity Act, 2003 and the Rural Electrification Policy envisaged decentralised management of distribution and also generation in rural areas. This has yet to happen in any significant manner. The real reason why this has not happened is that a commercially viable business model for this purpose has not been put in place. This has not been feasible in the absence of shadow prices which reflect real costs and therefore real subsidies. The utilities do not have the internal incentives for this. The Regulatory Commissions can get the real costs out in the public domain. Once that happens it would be much easier to evolve a workable business model for a franchisee. Through the efficiency gains which a franchisee can bring about it should, in theory, be possible to actually lower the subsidy burden on the government.

Private sector participation in distribution has been seen as the key to power sector reforms. The mega power policy crafted in the mid-nineties sought to make the cheaper power from the mega power projects with the benefits of import/excise duty exemption available only to those states which privatised electricity distribution in their cities with population of over a million. The MoU on reforms with Karnataka actually had a target date for the first privatisation of distribution. After the privatisation of distribution in Delhi there was, however, a period where both potential private sector participants as well as state governments felt the need to wait and see how this experience went. Delhi has worked well and is now finally being seen to have done so. With more recent decisions on franchising arrangements for urban areas, the time appears to be getting opportune for a fresh wave of efforts at private sector participation in distribution. While this would remain essentially a political decision in state governments, the SERCs can facilitate the process. Like the process for tariff based bidding whose outcomes the Regulatory Commissions accept, an acceptable transparent process for inducting private sector participation in distribution needs to be evolved and put in place. It would be good if SERCs through the Forum of Regulators (FORs) addressed the issue. Needless to add the distribution business must appear profitable enough to attract private investments.The underlying premise is that the efficiency gains through private sector participation would be large enough for both private profit and enhanced consumer welfare. One should expect private sector participation in distribution to be the next positive wave in the power sector.

For all the difficult challenges and opportunities that lie ahead before the Regulatory Commissions, their capacities to discharge their responsibilities need to be adequately enhanced. The commissions have not hesitated in using outside consultants. Nevertheless, they also need improved in-house capabilities. After all consultants have to be given specific tasks and their output has to be understood and judged. This cannot be done without a capable staff. Emoluments, perquisites, housing, career growth opportunities, in-service training and evaluation are all complex issues and not amenable to easy quick fixes and solutions. But these issues need to be addressed without any further delay because without doing so it would be unrealistic to expect the commissions to be able to do justice to their onerous responsibilities. In this context it is also necessary to put in place institutional capacities where formal courses and training programmes in this new, complex and rapidly evolving field can be conducted. Given the size and diversity of India more than one institution needs to be nurtured for this purpose. This is a major gap today.

One issue which does come up often is the choice of persons as members and chairmen of the commissions. The selection process as provided in the Electricity Act is robust enough. Given the paucity of talent and outside expertise in this new area choices have tended to be from the utilities and the government. As staff within the commissions matures they would become a natural source of talent for consideration. As professional institutions generate expertise and gain experience they would also become a potential source of talent. As time passes there is likely to be greater variety in the backgrounds of the members and chairmen of the commissions.

There is one important area where commissions need to pay more attention right now and that is public outreach and communication. In marked departure from tradition and in deference to the realities of the new media age, the chairman of the Federal Reserve, the US central bank, has decided to meet the media every quarter. The commissions being relatively young institutions are not so well known and are yet to be perceived by common consumers as the custodians and guardians of their interests. The fact is that it is this constituency which can make the task of the commissions easier as they move into unchartered territory and try to create new paradigms for the development of the power sector in the country. India today is not only far more accepting of change and innovation but it is also demanding it and that too with impatience. The ERCs have an historic opportunity and privilege today for transforming India by 2020.